COMPOSING SCIENCE

COMPOSING SCIENCE

A Facilitator's Guide to Writing in the Science Classroom

Leslie Atkins Elliott,
Kim Jaxon, and
Irene Salter

Foreword by Tom Fox

Published simultaneously by Teachers College Press, 1234 Amsterdam Avenue, New York, NY 10027 and by the National Writing Project, 2105 Bancroft Way, Berkeley, CA 94720-1042

This material is based upon work supported by the National Science Foundation under Grant No. 1140860. Any opinions, findings, and conclusions or recommendations expressed in this material are those of the authors and do not necessarily reflect the views of the National Science Foundation.

Figure 1.1 is reproduced by kind permission of Cambridge University Library.
Figure 1.3 is reproduced courtesy of the Ava Helen and Linus Pauling Papers, Special Collections & Archives Research Center, Oregon State University.
Figure 1.4 is reproduced courtesy of the Wellcome Trust, London. Licensed according to the Creative Commons Attribution Only license CC BY 4.0
Photographers: Figures 1.5 and 3.10 by David Williams; Figures 3.2–3.4, 3.6–3.9 by Rachel Boyd

Cover design by Holly Grundon / BHG Graphic Design.
Cover photos by Leslie Atkins Elliott.

Library of Congress Cataloging-in-Publication Data

Names: Elliott, Leslie Atkins, 1976- author. | Jaxon, Kim, author. | Salter, Irene, author.
Title: Composing science : a facilitator's guide to writing in the science classroom / Leslie Atkins Elliott, Kim Jaxon, Irene Salter ; foreword by Tom Fox.
Description: New York, NY : Teachers College Press, [2016] | Includes bibliographical references and index.
Identifiers: LCCN 2016029108 (print) | LCCN 2016044304 (ebook)
ISBN 9780807758069 (pbk. :acid-free paper) | ISBN 9780807775141 (ebook)
Subjects: LCSH: Technical writing—Study and teaching (Higher) | Academic writing—Study and teaching (Higher) | Science—Study and teaching (Higher) | English language—Rhetoric—Study and teaching (Higher)
Classification: LCC T11 .E39 2016 (print) | LCC T11 (ebook) | DDC 808.06/65--dc23
LC record available at https://lccn.loc.gov/2016029108

ISBN 978-0-8077-5806-9 (paper)
ISBN 978-0-8077-7514-1 (ebook)

Printed on acid-free paper
Manufactured in the United States of America

24 23 22 21 20 19 18 17 8 7 6 5 4 3 2 1

For our children:
Kate Button Elliott, Carolyn Jing Salter, Owen Ming Salter,
Ashley Michelle Jeffries, and Nickolas Patrick Penning

Contents

Foreword

Science has a reputation for laws that predictably describe such phenomena as thermodynamics, motion, and gravity. Laws give science the appearance of certainty, of solidity, especially when compared with disciplines in the humanities or the social sciences. The authors of *Composing Science* acknowledge this reputation, but argue that science is motivated by, deeply engaged in, and captivated by *uncertainty*. Understanding already worked-out laws and principles does not represent the work of science, but of science in the rear-view mirror.

Composing Science, in a remarkably consistent way, examines scenes of students writing about uncertainty. Students wrestle not with memorizing the laws or facts, but with using writing to grapple with puzzling observations, incongruous theories, and competing explanations. What emerges from uncertainty is curiosity—an engaged sense of inquiry that motivates innovation and discovery—the real work of science-in-the-making. In engaging in this work, students are driven to use writing to find theories, explanations, and definitions that ameliorate their uncertainty. Writing is ubiquitous and broadly defined in *Composing Science*. Informal and predominantly individual notebooks lead to collaborative diagrams giving way to more formal revised definitions and final papers. Writing plays a central role in students' figuring out questions and puzzlements. Staging scenes of uncertainty that result in genuine curiosity would be enough to make a great book on teaching science writing. But the pedagogy described in *Composing Science* doesn't only recapture the sense of the uncertainty of discovery, it also articulates and examines the social and collaborative writing practices that science uses to produce knowledge and reduce uncertainty.

One of the many virtues of *Composing Science* is the sophisticated examination of these practices, especially the mundane and material: notebooks, whiteboards, diagrams, definitions, and so forth. In the chapter on whiteboards, for instance, the authors take up one of the many informal practices of science writing. Whiteboards, placed strategically for use with small groups of students, are a technology that students use to represent knowledge-so-far, a temporary registry of their learning. Using a whiteboard, as the authors explain, requires the group to accommodate a variety of understandings at their table and devise ways to make those understandings publicly intelligible beyond the audience of their table. These in-progress-yet-public documents of knowledge, shared around the room, give instructors a clear vision of where the students are, and an opportunity to question, compare, probe, and critique students' nascent theories. Whiteboards, as the authors state, are "shared, debated, revised, and refined." This practice, like all the practices discussed in *Composing Science*, doesn't simply replicate a form or genre of writing (e.g., using a model to teach how scientists write on their whiteboards and asking students to mimic it), but instead replicates the work of knowledge discovery that is facilitated by a practice that scientists use.

One of the student quotations that provide epigraphs for each chapter is particularly germane to the idea of knowledge discovery. It is from Chapter 7 on definitions: "But we just made it up. It's not like scientists just made it up. Wait. Scientists just made it up!" In the dynamic space of uncertainty created in the classrooms described in *Composing Science*, the instructor recognizes the need to stop periodically, work on precisely articulating what students know, revising and refining their knowledge collectively in what the authors call "summative" work. Getting to a summative point takes a great deal of writing where students, individually and in groups, carefully have defined particular phenomena and then assessed and revised their definitions in light of contrasts and conflicts with other groups. By doing so, students "are more likely to understand that these ideas are complex, nuanced, socially constructed over time, and embedded in a much larger process of model building and experimental evidence." *Composing Science* demonstrates in detail how to engage students in the dynamic-work-in-progress of science, rather than memorizing or simply replicating the work that science has already done.

The present participle in the title conveys a principle behind the pedagogy in *Composing Science*. Without question, teachers of science will find this book inspirational and useful, college teachers for sure, but also teachers up and down the curriculum. For the National Writing Project audience, such work elaborates familiar principles of teaching writing in a new context, extending and expanding our work. I found myself, a teacher in the humanities, wanting to be a student in the class described in this book, to write about discovering properties of light, describing color with precision, and engaging in the same questions that the students were exploring. Then I found myself imagining classrooms up and down and across the curriculum, organized around tables with whiteboards, staging scenes of uncertainty around language, writing, history, math, social studies, or any other discipline. Beyond science teaching, the authors of this book have captured in beautiful detail important principles of excellent teaching: Uncertainty drives curiosity; iterative practices of writing produce knowledge; students' discoveries, respectfully responded to, are an opportunity to support ever-more precise claims, to design pedagogy that replicates the work, but not necessarily the form, of academics and other professional learners.

—Tom Fox, director, Site Development, National Writing Project

Acknowledgments

As we argue in this book, writing does not happen outside of a social context. This is certainly true in terms of the support we received throughout our grant-writing, teaching, research, and book-writing process. We are incredibly grateful to our families, in particular, Richard Elliott, Jeff Jaxon, and Jason Salter. And we are indebted to our students and colleagues for their generosity of time, ideas, and insights.

We would like to give special thanks to all the students in our scientific inquiry courses whose ideas made this book possible. We'd also like to thank the student assistants—Rachel Boyd, Tony DeCasper, Andrew Lerner, and Vanessa Quevedo—who lesson planned, read student work, and supported our data collection. And Danielle Astengo, whose editing support was invaluable.

Introduction

> My feelings on writing about science have changed. I've always thought about writing for science as writing hypothes[e]s. I loved being able to write things in my own words and things that make sense to me. —Deanna

As teachers, the position that we take throughout our courses and in this book is that we should be engaging students in writing assignments that replicate the roles that writing plays in scientific disciplines, rather than replicating the particular structures characteristic of that writing. That is, in all fields, we use writing to have, remember, share, vet, challenge, and stabilize ideas. Over time, stylistic conventions have emerged that serve to facilitate this work: bound and paginated lab notebooks, abstracts, ways of referring to and including others' work, choices to use first or third person, discussion of methods, embedding figures, 3' x 4' posters, and so on. But much of the focus of writing instruction—particularly within the scientific disciplines—prioritizes the reproduction of these conventions rather than writing that generates, shares, challenges, or builds on the ideas of a community of scholars. And so while we believe that explicit attention to academic writing is a critical part of science teaching, our writing assignments perhaps are not what first come to mind when faculty think of a "writing-intensive" course in science. Students' writing does not, at first glance, look like a journal article or conference proceeding, but instead does the work that science writing does—developing and disseminating ideas.

This book is written for science faculty who are asked to teach "writing" in their courses, but aren't sure why or how to do so. We begin with a discussion of why science faculty are being asked to teach writing, and why their English faculty colleagues alone are not able to meet that need. We then discuss what writing in science looks like, and how that might play out in science classrooms from elementary school through university.

WRITING INSTRUCTION: WHY ENGLISH FACULTY ALONE CAN'T TEACH ACADEMIC WRITING

In K–12 settings, writing instruction, including elements like spelling, grammar, and organization of an argument, has long been part of instruction in English/language arts classes. Other classes draw on those skills but, in general, do not explicitly provide instruction in writing. Similarly, on most college campuses, particularly in the United States, incoming students are required to take some variation of "freshman comp," a composition course that prepares them to write at a college level. Like the K–12 English teacher, who prepares students for writing in other domains, college instructors, it is imagined, will introduce freshmen to resources they will be expected to use throughout their college career: the university library, search engines, and style guides; they will work with students on common practices in academic writing: how to embed quotations, include citations, create an abstract, and draw from the literature without plagiarizing; they will help students in crafting and supporting claims, shaping these ideas into sentences, paragraphs, and a well-sequenced paper devoid of grammatical errors. Armed with these resources and skills, freshmen will progress into their majors, prepared to craft papers that argue persuasively for their ideas, to use established conventions, and to draw on the literature of their chosen fields and on their own findings. The professors within each major, in this scenario, help students develop the content knowledge that allows them to populate their writing with ideas from the discipline. In this model, science faculty, in partnership with English faculty colleagues, prepare students who understand science and can write skillfully about those ideas.

There are several assumptions underlying this approach to writing instruction—assumptions that research in composition instruction has called into question. Among these are (1) that academic writing is a skill that can be taught and mastered independently of any particular discipline's knowledge and practices; (2) that the "term paper" or "five-paragraph essay" written for a composition course is characteristic of academic writing, so that mastering this genre will acquaint students with how knowledge is produced and prepare them for engaging with and producing this knowledge; and (3) that academic writing itself, whether it is the five-paragraph essay employed in secondary education or journal articles and book chapters characteristic of professors' work, is an important focus of our instruction.

This first assumption is widespread: We often describe students as good writers (as opposed to the more specific—and accurate—descriptions: a good poet, a careful note-taker, a persuasive editorialist, a brilliant comic strip author, an elegant proof constructor); we develop tests for K–12 students to assess writing abilities; we sort students into courses and programs based on that ability. And while there certainly are bits of knowledge that transfer from one context to another (e.g., spelling, some grammatical constructions) and strategies that, when learned in one area, may be helpful for another (e.g., analyzing good writing to improve your own, engaging in peer feedback and review), it is also true that much of what makes someone a "good writer" is not universal (Adler-Kassner & Wardle, 2015; Petraglia, 1995; Russell, 1991). Writing is an activity that is used by communities for a variety of purposes: What counts as "good writing" in one community may not represent good writing in another.

As one example, one of the authors—Kim, a faculty member in an English department—was asked to contribute a paper and present a poster at a physics education conference; these would introduce ideas from composition to physics professors. When she was crafting her paper, a range of "writing" problems surfaced. She consulted the style guide for how to embed lengthy quotes, but there was no information on how to do this, nor could she find examples in published works. The contrast to her own field, in which extended arguments are presented and built on or challenged as an expected practice (and often constitute the bulk of a claim), was striking. When told she would present a poster, Kim, unaware of conventions and software that are commonplace for presenting work in science, asked, "Do I need to buy a trifold poster board?" Similarly foreign were ways of using data, conventions around citations (e.g., referring to authors by last name only), and the relatively short (four-) page limit. And, of course, without a background in physics education, she had difficulty constructing a narrative for the paper: What kinds of arguments exist in the field? How should she position the work from her field to speak to a physics audience? Academic papers (indeed, all texts) stand in conversation with others, and a large part of learning to "write scientifically" is not only the skillful use of the conventions of scientific writing, but understanding the ongoing conversation and how to enter that conversation. So Kim—a fully competent writer in her own field (composition, no less)—would not be characterized as a competent writer by a faculty member in science. Indeed, the reviews of her paper primarily critiqued the presentation of the ideas, rather than the ideas themselves.

Of course, there are characteristics of academic writing shared among the disciplines. Loosely speaking, what unites the work of university faculty is the production of knowledge, based in methods characteristic of their disciplines, and the sharing of that knowledge, usually through peer-reviewed journal articles or books. While Kim was not familiar with the particulars of how to write for a physics conference, she understood that her paper should fit within a particular style and conversation, that it should build on prior work, and that it should make claims and support those claims. And so it seems reasonable that a freshman comp course might engage students in writing research papers so that they might better understand the ways in which knowledge is produced and shared. Students might then be prepared to do that work within their major—albeit using different conventions. However, this would require that the introductory composition course afford students an opportunity to engage in research. With some exceptions, most freshman composition courses are not structured as research courses, and so students' assignments become, at best, a literature review, an opinion piece, or an overview of rhetorical modes, rather than representing any novel research on the students' part. In this way, the work of freshman comp only very weakly approximates the kinds of writing characteristic of academics' work. And so the second assumption behind traditional "composition" courses—that the writing from these courses is, in some way, representative of academic writing in general—is rarely the case (Russell, 1995). And even when it is the case, faculty in science are still responsible for teaching writing particular to the sciences, and they cannot expect that students will know conventions around writing in scientific disciplines.

Finally, if we suspend skepticism regarding the first two assumptions and imagine that there are some commonalities in academic writing that can be taught in composition courses, and that courses are structured in a way to allow that, a broader question remains: Does replication of academic writing—that is, assignments that ask students to replicate the conventions of writing that academic professionals do—best address students' needs? Most students, particularly nonscience majors, will have few occasions to read or write scientific research articles, and an awareness of their conventions will be of little use; but all students will be consumers of this research, and we hope that they will be able to understand how the ideas in those articles were constructed, debated, and vetted. Our courses should give them an opportunity to engage in and understand that process and the role of writing in it.

Recent changes to curricula in both K–12 and university settings have recognized this. In university settings, "writing across the curriculum" movements, now widespread, call on faculty in all disciplines to engage students in writing characteristic of their particular discipline. In K–12, the Common Core State Standards for English Language Arts & Literacy in History/Social Studies, Science, and Technical Subjects (Common Core) (Common Core State Standards Initiative, 2010) are calling for more attention to reading and writing in academic disciplines outside of English, and the Next Generation Science Standards (NGSS) call for teachers to attend to the role of scientific practices—like reading and writing—in developing scientific ideas.

Below we discuss in more depth what we mean by "science writing" (which we construe broadly) and describe the structure of this book and how to use it.

WHAT SCIENTISTS WRITE

So how do scientists use writing? We scribble on chalkboards, draw diagrams, annotate graphs and photos, jot down ideas in notebooks, send emails, scrawl notes in the margins of journal articles, write grant proposals, review proposals and papers, draft conference proceedings, put together presentations, and, ultimately, tidy up all of this work into a publishable journal article. This all takes place as part of the gradual, iterative process of developing scientific ideas in a research community. In science—as in all fields—writing is not just a way to communicate what was learned, it is an essential tool for developing and vetting new ideas. Sociologists of science have gone so far as to propose that writing is not simply a supporting activity—a tool for producing scientific ideas—but can be considered the central activity of scientists. A nonscientist foreign observer in a research lab "soon recognizes that all the scientists and technicians in the lab write in some fashion, and that few activities in the lab are not connected to some sort of transcription or inscription." This observer finds the laboratory to consist of a "strange tribe" of "compulsive and manic writers . . . who spend the greatest part of their day coding, marking, altering, correcting, reading, and writing" (Latour & Woolgar, 1979, pp. 48–49). In this sense, writing in science courses should not simply present a means of assessing students' writing abilities and content knowledge, but should be a central feature in developing scientific practices and knowledge.

It's often novel, but not too difficult, for science faculty to consider these various and informal forms of writing as what is meant by "writing in science" and to include these as the kind of writing instruction they teach and assess in their courses. What has been harder for us as we have considered our students' writing is to construe writing even more broadly, so that "writing" encompasses not just written words (wherever they may appear), but the simulations, models, diagrams, equations, and presentations that are composed alongside—or in place of—those words.

Researchers, however, have long noted that the "role" of text is assumed not only by written words, but also by a host of discipline-specific forms: models, diagrams, equations, and so on. A focus on text alone misses many of the ways in which we construct and share ideas, particularly within scientific disciplines. And so in our "writing-intensive" science courses, we should consider not only students' written texts, but also, more broadly, the diagrams, presentations, simulations, and physical models they compose as they engage in scientific inquiry. The field of composition and literacy studies now includes these kinds of compositions in what it considers "writing," and when "writing across the curriculum" scholars call on faculty to teach writing in all disciplines, it is precisely these multimodal compositions that they are hoping faculty will include in their courses (Shipka, 2011).

STUDENTS' SCIENTIFIC WRITING

K–12 students and undergraduates, however, rarely are introduced to these varied writing practices, at least not formally. Instead, literature reviews (a.k.a. "research papers"), school-based lab reports, and assignments modeled on the professional journal article are often the only forms of what it means to write scientifically; they are what typically "count" for writing in K–12 and undergraduate science. Style guides, from either journals or course-specific rubrics, determine how students will communicate scientific ideas. While the journal article is a core genre for academic and professional scientists, it is far from the only—or even the most common—writing that scientists do and may not be the best introduction to crafting, vetting, and communicating scientific ideas. This is particularly true for nonscience majors or those who do not plan on careers as research scientists. For them, as outsiders, the specific idiosyncrasies of journal publications, designed to facilitate communication and clarity within a specific field, often do the opposite and may distract students from attempts at clarity. The potential benefits—a familiarity with conventions specific to the discipline—are of scant value to these students.

We knew that we wanted our students (primarily undergraduate nonscience majors interested in K–8 teaching careers) to use writing in the way that scientists do—that is, to use writing in formative ways that push them to generate, persuade, critique, challenge, and defend their own developing scientific ideas in conversation with their classmates. We also wanted to include summative writing assignments that required this work to be formally presented, where students would communicate more conclusive findings that emerged from their inquiry.

We found that in order for our students to *write* scientifically, they needed a course that engaged them in *doing* science. "Doing science" is used commonly in describing novel teaching methods, and it can be construed in many ways—in some contexts, it means treating students as peers and setting high standards for their content knowledge; in others, it means having students assist in actual research labs, where they come to understand how current scientific ideas are constructed as they are apprenticed into research; in still others, it means expecting students to master scientific skills, such as control of variables, reasoning from evidence, or following scientific protocols. For the purposes of teaching our undergraduate nonscience majors, we use "doing science" to mean that students are engaged in a social activity in which they construct models of phenomena, test those models, and construct and defend arguments for the models they have developed. We spend significant amounts of class time with students engaged in writing—jotting down notations in lab notebooks, diagramming on whiteboards, reviewing one another's work, and crafting clear descriptions and definitions. However, this writing is embedded in doing science, rather than as a separate task. Chapters 1–8 of this book are devoted to the techniques we use to teach these embedded writing tasks.

HOW TO USE THIS BOOK

The remainder of this book is designed to be more pragmatic than theoretical. It is meant to be *useful*, with vignettes from our classes, suggestions for things to try, and ideas about how to give meaningful feedback. Each chapter focuses on one aspect of scientific writing—calling attention to the role that this aspect serves in the scientific community—and provides suggested assignments and activities that engage students in doing that work, with examples from our courses. For example, peer review is a means by which scientists offer feedback on ideas and arguments; but when students are asked to provide feedback on a peer's paper, their first impulse is to comment on mechanics and grammar, rather than on identifying a paper's main arguments and the merits of those arguments with respect to the community's ideas. We offer suggestions for addressing that impulse. In the end, while the writing may not look like a typical peer review, the role that it plays in our classroom has strong parallels to the role played in the scientific community.

This book is divided into three parts. Part I, "Writing to Learn: Informal Writing Strategies That Generate New Scientific Ideas," addresses writing strategies that primarily emphasize the development of scientific ideas (e.g., notebooks, diagrams, reading, and interpreting others' work). This is the time when students collect data (make observations of seeds grown in the dark versus in the light), ask questions (why did the seeds in the dark grow at all?), and try out different ideas (it could be that plants can get energy from the soil, that light leaks in, or that the seeds had energy stored up inside them). We consider this time of raw observations, data gathering, and initial idea sharing the "write to learn" phase. Like their ideas at this stage, we accept that students' writing will be messy, personal, and half-baked—like an email you might share with a colleague or some hand-drawn notes you might discuss with a lab mate over coffee. We aren't yet interested in polished products or well-defended claims, but in whether the students are making sense of their observations, constructing and building on or challenging the ideas of others. The chapters in Part I offer suggestions for varied approaches to "writing to learn," such as student notebooks (Chapter 1), whiteboards (Chapter 2), diagrams (Chapter 3), ways to critique and build on the raw ideas of others (Chapter 4), readings (Chapter 5), and homework assignments (Chapter 6).

Part II, "Writing to Communicate: Formal Writing Strategies That Share, Critique, and Defend Scientific Ideas," tackles writing strategies that primarily emphasize sharing and defending scientific ideas (e.g., more formal assignments, definitions, and papers). Here, students have arrived at an idea that makes sense to them and use writing to communicate and justify that idea to others. This writing is more formal. There is a specific audience to consider—in our classroom, that audience is usually their peers and the instructor, but occasionally we have students write to one another or even to a class of 5th-graders. These assignments include concise definitions (Chapter 7) and final papers (Chapter 8).

Of course, the dichotomy between Part I and Part II is not clean—peer review is often a critical means of sharing and defending ideas, while students working on their definitions often find themselves revising their ideas and considering alternative explanations more seriously. A whiteboard of tentative ideas may be revised and refined to argue a claim. Students often have "aha" moments during exams, when ideas come together or a model is extended to a new phenomenon.

Finally, in Part III, "Adaptations: Bringing Writing Strategies into Different Settings," we examine how to use these ideas in other settings. The course where we teach writing in science is an open-inquiry, lab-based course in the sciences, taken primarily by nonmajors who are planning on teaching careers. As noted above, a central practice in our teaching is that students are the authors of their own ideas and these ideas are developed in a social context. This is obvious throughout our work, including the kinds of writing that we assign and the way that we assess that writing. We imagine that the ideas in this book will be most easily implemented in courses like our own: small lab courses that allow for extended interactions among students, freedom to explore scientific questions through open-ended investigations, and the foregrounding of students' ideas. We anticipate that many college and university faculty have some freedom in designing nonmajors' courses and can imagine ways to use similar structures in their courses. However, for developers of courses (often those for science majors) that must meet a long list of content goals, large-enrollment courses that allow for limited in-class interactions with faculty, and courses without labs, implementing these ideas will take some adapting. We offer a chapter of suggestions for those course designers. In addition, in our final chapter we offer some suggestions specifically for K–12 teachers. The approaches to writing we address in this book are consistent with standards in K–12 education: the Next Generation Science Standards (NGSS Lead States, 2013) and the Common Core State Standards for English Language Arts & Literacy in History/Social Studies, Science, and Technical Subjects (Common Core State Standards Initiative, 2010). The connection between our approach to teaching writing in science and the standards is clearly important for those who teach in K–12 settings, but we also encourage university faculty who do not work with K–12 audiences to consider the implications of these standards to college teaching and learning. The standards are drawn from extensive research on how students learn and how practitioners develop knowledge, and how to best engage students in developing content-area literacy for college and careers.

THE STRUCTURE OF THE CHAPTERS

The chapters in Parts I and II have a standard format. They each address a particular aspect of scientific writing: lab notebooks (Chapter 1), informal but public discussions at whiteboards (Chapter 2), diagrams (Chapter 3), conducting peer reviews (Chapter 4), reading journal articles (Chapter 5), working out ideas on one's own (homework) (Chapter 6), constructing definitions (Chapter 7), and writing a final manuscript (Chapter 8). For each topic, we begin with an introduction to that aspect of writing in science and discuss the role that it plays in developing scientific ideas. In some cases, we provide scripted ways of rehearsing the activity before you begin. These scripts aren't a step-by-step lesson plan per se, but are rather a routine you might establish so that it is easier to try something new. For each chapter, a step-by-step lesson plan is available in the online resources that accompany this book (see composingscience.com).

Next come examples from our classroom that illustrate how this strategy is "taken up" by students. In other words: What do students actually do with their notebooks or with whiteboards? What does that look like in the classroom and how does it play out? How is the work of students related to the scientific practices that we hope to cultivate? Here we offer examples of our own students' work, drawn from the course Scientific Inquiry. The students we describe are majoring in Liberal Studies, a program designed for future elementary teachers; in most cases, these are students who do not think of themselves as particularly skilled at science or interested in scientific research.

Much of the writing in our class is different from what students have experienced in science classes, and it takes a bit of scaffolding to orient them to these new routines. As we have discussed this course with colleagues, they have had a range of questions related to the routines. So we next discuss the "challenges," for students and faculty, that we have encountered and what we have done in response.

The next section, "feedback and grading," discusses ways to provide both formative feedback to your students and, more summatively, give them a grade for their work. To us, feedback and grading serve different roles. Feedback is akin to what might happen when you circulate a draft to colleagues or share ideas at a conference or other settings wherein you discuss your ideas prior to a more summative and formal journal submission. The feedback that students receive from faculty and peers is meant to help students develop their ideas further, consider other points of view, and work on clarity in their representations. The feedback that students offer their peers is critical, too, as it is a way in which they deeply consider others' ideas, alternative representations, and data they might not have noticed or considered. In this way, much of the feedback on student writing is performed by students; this mirrors scientific work, with the added benefit of reducing the instructor's workload. (Perhaps the most helpful piece of advice we received in talking with colleagues in composition was: "You don't have to touch every piece of writing. If you have time to give feedback on every piece of writing, your students are not writing enough.")

In contrast to feedback, grading is done by faculty and assigns a number to a student's writing, something that is qualitatively different from giving detailed feedback as part of the development of an idea. Grading is comparable to an editor making a decision on your paper, the NIH assigning a score to your grant, or a conference organizer deciding whether to include your presentation in the proceedings.

Finally, we summarize each chapter with some "take-home messages." Here we summarize the key themes from the chapter and provide concrete information that you can use right away.

MECHANICS AND GRAMMAR

You can anticipate that your students will struggle with grammar, mechanics, and other conventions as they learn to write in a new field. This does not mean that your students have not learned, for example, how to write persuasively or what constitutes a run-on sentence. Instead, as noted in Appendix A of the Common Core (2010), it often indicates that your students are learning a new and complex task:

> Grammar and usage development in children and in adults rarely follows a linear path. Native speakers and language learners often begin making new errors and seem to lose their mastery of particular grammatical structures or print conventions as they learn new, more complex grammatical structures or new usages of English, such as in college-level persuasive essays (Bardovi-Harlig, 2000; Bartholomae, 1980; DeVilliers & DeVilliers, 1973; Shaughnessy, 1979). These errors are often signs of language development as learners synthesize new grammatical and usage knowledge with their current knowledge. Thus, students will often need to return to the same grammar topic in greater complexity as they move through K–12 schooling and as they increase the range and complexity of the texts and communicative contexts in which they read and write.

However, as should be clear, this is not a book that suggests explicit methods of instruction in grammar and mechanics. And in fact, most writers, both newcomers and old-timers in any discipline, will never "master" the conventions of language, particularly since language is a living thing.

What we *can* do as instructors is give students plenty of opportunities to try out the writing practices in our field, share in the responsibility for teaching writing mechanics (especially noting when a particular convention is part of our academic genre), and, perhaps most important, model expectations about reading their own writing and the writing of others with care. As the field of composition and literacy studies has demonstrated, learning the mechanics of writing does not precede writing; we learn to edit our own prose over and over again as we work through complicated ideas. As students' ideas and arguments become clearer, their writing becomes clearer.

Part I

WRITING TO LEARN

Informal Writing Strategies That Generate New Scientific Ideas

Student Notebooks

> We are judged for *what* we know by the homework and exams, but the journals give us a chance to get a grade for our ideas and being able to develop them. If we are only graded on our exams, we are simply regurgitating the information we learn back to you. By writing down our ideas and how they develop, we are much more likely to retain this knowledge. We have a chance to write it down, review what we wrote, correct our misconceptions, etc. —Dawn

Notebooks are a central feature in professional scientists' laboratories and in science classrooms. In some cases, student notebooks approximate those used in professional academic, government, and industry settings—with requirements to sign every page, never erase, and get one's notebook witnessed by peers (Roberson & Lankford, 2010). In other cases, notebooks are teacher-centered and guided by the "scientific method" or other heuristics (Keys, Hand, Prain, & Collins, 1999; Klentschy, 2005). We personally recall from our own science courses very particular requirements that were established in order to ensure that we wrote down the information needed to prepare a polished lab report later on: our hypothesis, procedures, and data that led to conclusions our instructor anticipated in advance. Ultimately, we and our classmates submitted nearly identical notebooks, with accurate data carefully recorded, that were consistent with the goals of the lesson. Unfortunately, research suggests that students often do not see the benefits and utility of using notebooks when given a choice to use one or not (Garcia-Mila & Andersen, 2007); and when notebooks are required, they often are mechanically completed to respond to teacher prompts, with little evidence of student reflection or understanding (Baxter, Bass, & Glaser, 2001).

This stands in contrast to the ways in which scientists use notebooks. While many scientists anticipate in advance the kinds of papers they might write, and record data accordingly, a notebook also serves as a personal diary of one's scientific life—a place to think; to express one's individual personality as a scientist; to record personal thoughts, questions, and mistakes; and ultimately to help the scientist reconstruct a story of how ideas have changed and why. The science notebooks of practicing scientists are anything but mechanical and identical, even among scientists working in the same laboratory. John Dewey, writing in 1920, recognized that "to require [students] to use exactly the same plan may make the checking of notebooks more easy and their appearance more satisfactory, but it stifles the pupil's originality and prevents him from discovering and correcting his own faults" (as reported in Russell, 1991, p. 96).

The notebooks of scientists vary greatly from one person to another, from one lab to another, between academia and industry, and between scientific disciplines. What ties them all together is a dual role for notebooks—as something convenient, useful, and personally meaningful to the individual scientist going about his or her work, and also as something that keeps that individual scientist accountable to the greater scientific community (Shankar, 2007, 2009). As Shankar (2009) writes:

> Scientific records, particularly in academic science, occupy a curious niche. On the face of things, they are organizational documents that fulfill the expectations of the scientific community, the immediate laboratory group, the university, and the individual scientist. But in that last role, they are also profoundly personal documents, a diary of learning, expertise, and meaning. They are created by human beings, but act back upon them in complex ways, documenting the interactions and intersections of memory, knowledge, space, and time. (p. 159)

Consider this dual role for notebooks as you examine the notebooks of Charles Darwin (Figure 1.1), the Manhattan Project (Figure 1.2), Linus Pauling (Figure 1.3), and Marie Curie (Figure 1.4). It is not important to understand the work each scientist did, but rather to notice the ways in which they took notes and organized information. As with our students, you might consider the following as you read the notebooks of scientists: What does the page look like? What is written down? What is the style of writing—personal, objective, colloquial, and so forth? What might be the purpose of a scientist's research notebook?

These notebooks are strikingly different. Some pages are numbered and dated (like Pauling's), others have only a number (like Darwin's), or only a date (like Curie's). The Manhattan Project notebook is written in pencil; the others are in pen. In all cases, however, they serve as a public record of scientific activity—a history of what the scientists have done and how they make sense of the data. In some cases, this documentation may be in anticipation of future patents; in others, it is written to provide documentation for other scientists. For instance, the page from the Manhattan Project notebook (Figure 1.2) contains a highly organized table of the data. The experiment under consideration documented how the nuclear reactor responded to moving cadmium control rods that prevented the chain reaction from spiraling out of control. Since it was part of a classified project, it is not likely that the authors anticipated scrutiny by a larger scientific community; rather, it is more likely that the data were critical for a scientific team, with

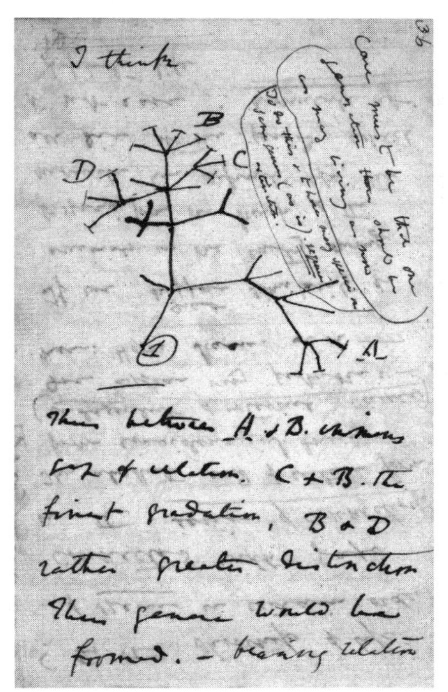

Figure 1.1. Image from Charles Darwin's Notebook

Courtesy of the Cambridge University Library

Figure 1.2. Image of Manhattan Project Notebook

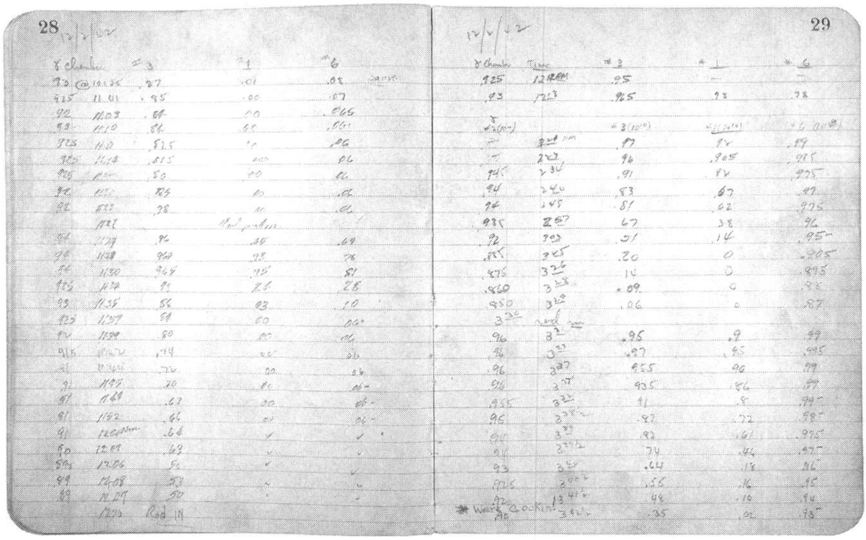

From the Records of the Atomic Energy Commission;
Record Group 326; National Archives

Figure 1.3. Image from Linus Pauling's Notebook

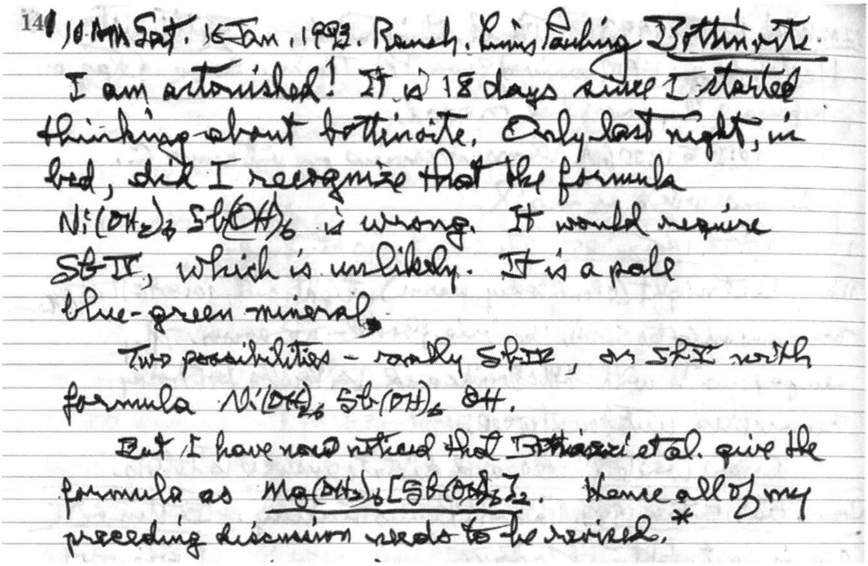

14 10 AM Sat. 16 Jan. 1993. Ranch. Linus Pauling Bottinoite.
I am astonished! It is 18 days since I started thinking about bottinoite. Only last night, in bed, did I recognize that the formula $Ni(OH_2)_6 Sb(OH)_6$ is wrong. It would require Sb IV, which is unlikely. It is a pale blue-green mineral.

Two possibilities – really Sb III, or Sb V with formula $Ni(OH_2)_6 Sb(OH)_6 OH$.

But I have now noticed that Bottiazzi et al. give the formula as $Mg(OH_2)_6[Sb(OH)_6]_2$. Hence all of my preceding discussion needs to be revised.*

Courtesy of Ava Helen and Linus Pauling Papers, Oregon State University Libraries

Figure 1.4. Image from Marie Curie's Notebook over a Span of 4 Days

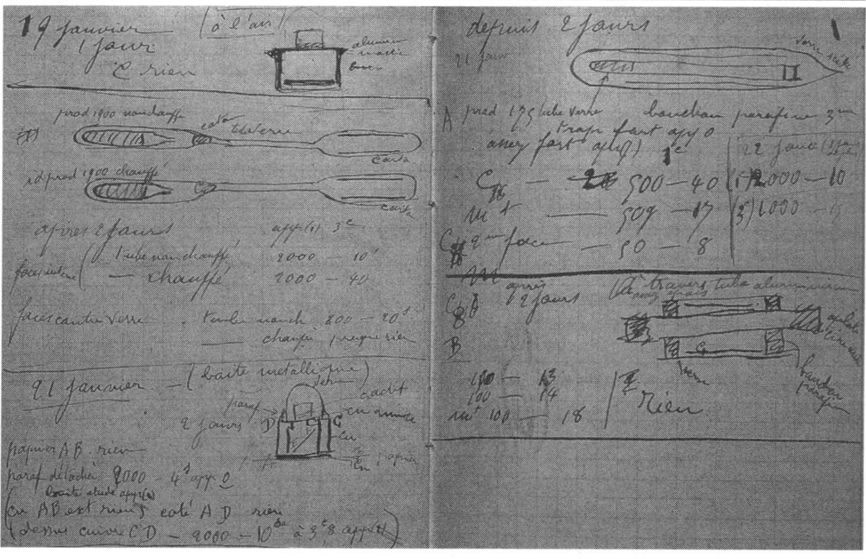

From the Wellcome Trust, London.

this experiment but one piece of a much larger puzzle. As such, this lab notebook must provide detailed, complete information about the investigations.

On the other hand, in most cases, a notebook is also a very personal artifact; it is a diary of one's scientific life. It is a place to wonder, to sketch, and to express thoughts, questions, and mistakes as one works. These notes may be creative and messy. They may be jotted down after a middle-of-the-night inspiration like Linus Pauling's realization that the formula for bottinoite that he initially had proposed was wrong (Figure 1.3). They may be written in shorthand in the margins like the "We're cooking!" notation at the bottom of page 29 in the Manhattan Project notebook.

To develop notebooks that function in the ways that notebooks function for practicing scientists, we ask our students to examine pages from several scientists' notebooks and to use these as a model for their own notebooks. We give students time to examine the scientists' notebook pages in groups, then ask them to consider the questions we asked you at the start of this chapter about the organization, style, content, and purpose of a science notebook.

Students note the diversity of the notebooks and observe that they may include notes, drawings, equations, questions, ideas, dates, tables, and more. They note that some are neat while others are messy and hard to read. More often than not, the scientists don't use complete sentences, preferring personal notes, lists, or shorthand.

Ultimately the class is charged with turning these observations and ideas into a rubric to which they hold themselves accountable. Students wrestle with whether titles and a conclusion are required for every entry, what counts as "relevant information" (and how that goes way beyond materials, methods, and data), whether pages should be numbered, and what it really means to justify ideas. The instructors then take the class consensus and shape it into a grading rubric.

Every semester that we've set up notebooks this way, students have identified the two major roles discussed earlier. They observe how the notebook is both personally meaningful to the scientist and publicly useful to the scientific community. Thus, we often divide the rubric into two sections: a "personal" section that acknowledges that there will be mistakes, dead ends, individualization, creativity, the telling of the scientific story, and messiness (no two lab notebooks should look the same); and a "public" section that emphasizes clarity, organization, detail, precision, and completeness.

TAKEN UP

When we set up notebooks in this way, students quickly adopt a scientific approach to notebooks in that they make inferences, show evidence of metacognition, use the notebooks to better communicate with one another, and utilize diagrams (Atkins & Salter, 2011). Students come to see their notebooks as a way to learn and not simply as a way to record what is already known.

Our findings mirror those of other educators who introduce notebooks in similar ways. Both Lunsford, Melear, Roth, Perkins, and Hickok (2007) and Morrison (2008) introduced preservice teachers to science notebooks by first having them examine scientists' notebooks and then observing changes over time as they gained familiarity with using notebooks during inquiry. Students in Morrison's study gradually moved away from a teacher-centered view of notebooks as a course requirement and shifted toward a more personal, dynamic, student-centered view. Lunsford et al. documented a transition toward deeper scientific thinking in the number and types of inscriptions students used over the course of the semester.

Below, we share five examples from our students' notebooks that highlight what our students do with science notebooks and how that relates to the scientific practices we hope to cultivate. These are all taken from a course for undergraduate preservice K–8 teachers.

Jordan

One semester, when seeking to resolve a debate regarding beams of light, students examined the spot of light created by shining a flashlight down a tube. Noticing a hazy edge to the spot of light, students determined that light reflects off the walls of the tube. Ultimately, they decided that the reflection of a light ray off a nonreflective surface is like a "Koosh ball" sending rays out in all directions (what a scientist would call diffuse reflection). Students coined the term "seconds" to describe rays that have reflected off the walls once, "thirds" for rays that reflect twice, and so on; they named the reflection itself the "kush [*sic*] effect." As you read Jordan's notebook (Figure 1.5), notice how he has taken up the vocabulary coined by other groups as his group works to better understand the hazy edge on the spot of light.

One implication of the "kush effect" is that light is not actually confined by the tube to reach only one spot, as might be expected. Instead, rays can reflect off the inner walls of the tube to reach parts of the paper that aren't directly in front of the tube. Jordan describes the question his group hopes to answer: "We want to know why we saw a continuous fade of brightness in concentric rings from the initial pool." In the diagram, you can see the experiment they tried in an effort to account for their observations (they added a triangle of black tape halfway down a black paper tube), along with the observations and a modeling of the light.

Lab notebooks—both in our courses and more generally—often are used as a place to record questions, methods, data, and conclusions. In many courses, this content is segregated and carefully sequenced in accordance with the headings of a typical scientific research paper. Yet, as you

Figure 1.5. Jordan's Class Notebook

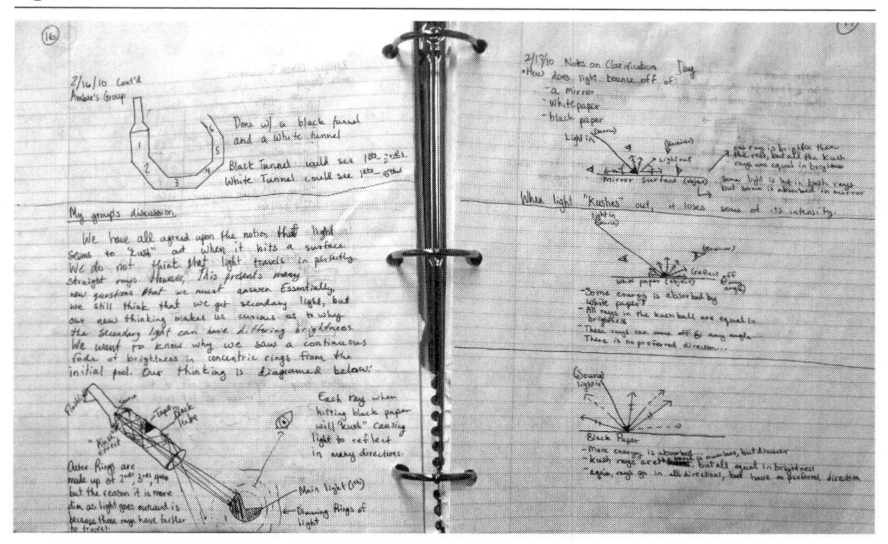

can see from the examples of famous scientists' notebooks, in the actual practice of science, the content is rarely compartmentalized and sequenced. Like Darwin, Jordan weaves together explanations, questions, methods, diagrams, and theory into a story that clearly does not follow the rigid format that is represented in scientific journals and that we authors recall from our grade school and university lab notebooks. One of our students, Megan, describes the lack of strict guidelines in this way: "The writing I've done in this class is so different [from] any writing I've ever done in the past because it was so free-flowing, especially in our notebooks. I was able to write for myself instead of a teacher/grade."

Jordan's notebook also exemplifies how new vocabulary is generated and adopted by students as they strive for clarity of understanding. Whereas many curricula "front end" definitions—students are given vocabulary with precise definitions early on in their investigations—our course encourages students to coin new terminology as it becomes necessary to do so in response to the demands of the observations or theoretical progress being made, and to anticipate modifications to that terminology as needed (Atkins & Salter, 2011). (See Chapter 7 for much more on definitions.)

Lauren and Tyson

The next two examples are drawn from a semester in which we were studying color and were using simple diffraction grating spectroscopes to visualize the colors in the light from different sources. Students had become interested in how colors mix. Some groups had been experimenting with mixing paints while others had been investigating colored filters. Tyson's group observed that stacking two filters on top of each other and looking through both together produced dramatically different results than when two overheads, each with a different-colored filter, were shined on the same screen. For example, a red filter stacked on a green filter lets no light through, so one sees black, while the projection from a red filter overlapping with a projection from a green filter gives the perception of yellow. Lauren and Tyson are now puzzling over how and why this might be.

As you read Lauren's notebook (Figure 1.6) and Tyson's notebook (Figure 1.7), notice how they weave questions, observations, models, and theories in the same way that Jordan does. Moreover, notice how the organization and format of Lauren's and Tyson's notebooks are different from each other (and also from Jordan's notebook) even as they puzzle over the same observations in the same classroom.

The individual personality of each student comes through in these two notebooks. Lauren's entire notebook (like her person) is very nonlinear in the way she organizes the page, drawing attention to the relationships between ideas, questions, and thoughts as if it were a concept map, rather than the narrative story that Jordan tells. Moreover, hers is very personal and sincere, with exclamations of "What the heck?!" and "Seriously, so weird!" just as Linus Pauling's notebook exclaims, "I'm astonished!" In contrast, Tyson's puzzlement is expressed more quietly.

With these puzzlements as a focal point, we can observe how Lauren and Tyson chronicle the development of their ideas over time. For instance, you can see how Tyson puzzles over what the combination of blue and yellow should be. Similar to Linus Pauling's entry (Figure 1.3) where he states: "Hence all of my preceding discussion needs to be revised," Tyson begins: "Never mind . . . Yellow (under a spectroscope) gives us D ROY G B. So everything I said today is wrong."

He comes to realize that stacking filters can give results similar to mixing paints and concludes: "paint and light are similar and react much in the same way." (Interestingly, he has yet to disentangle the two different ways to combine lights. Thus, what he crossed out actually is scientifically right—that mixing inks is not like combining the light of two projectors—yet he now realizes that combining paints does not obey a completely different set of rules than combining lights colored by filters.) Because notebooks become a space in which to reason out ideas, notebooks are a prime example of how students "write to learn."

Emma

While the notebook's main function is to support the construction and documentation of a student's scientific thinking, some students also use their

Figure 1.6. Lauren's Class Notebook

February 5, 2014

Stack green on yellow
what do we get?

Green filter blocks some & the red filter blocked the rest to be black

After trying multiple combos, our idea is that when we add darker filters togethers, we would get black

red filter & dark blue we should get black
BUT we didn't!! What the heck?! It was a maroon-y color

That's weird, we thought these 2 dark colors would filter all light & become black, but they didn't

Let's do more & see if we can understand!

So then let's try red with the light blue filters. This should still be a light color, won't totally block light

Results: BLACK
Seriously, so weird! Really interesting because this goes against our prior thought of adding two dark filters to get black, but it flipped this idea

Why does it react differently the second time around (with the light blue & red filter)?

Why not the same behavior as others?!

Figure 1.7. Tyson's Class Notebook

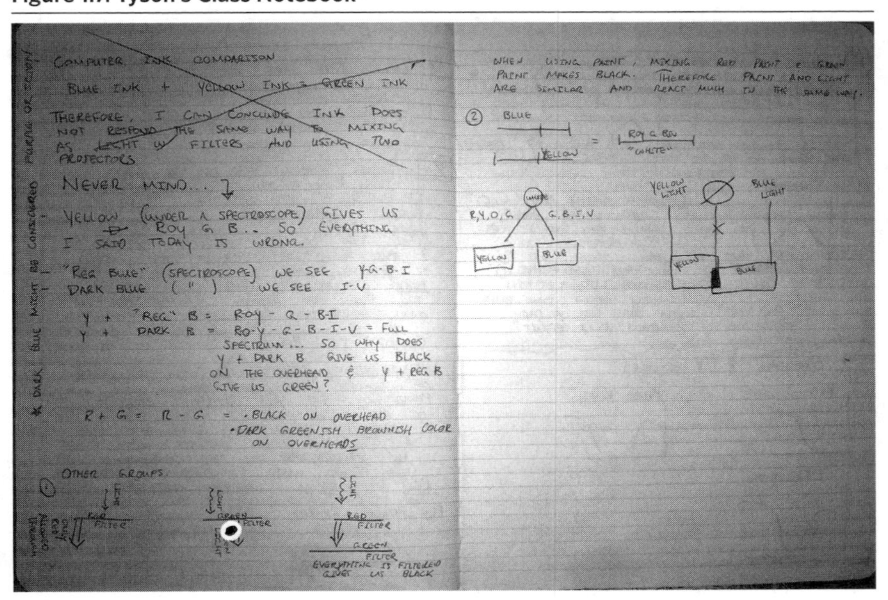

~~COMPUTER INK COMPARISON~~

~~BLUE INK + YELLOW INK = GREEN INK~~

~~THEREFORE, I CAN CONCLUDE INK DOES NOT RESPOND THE SAME WAY TO MIXING AS LIGHT W/ FILTERS AND USING TWO PROJECTORS~~

NEVER MIND...

- YELLOW (UNDER A SPECTROSCOPE) GIVES US ROY G B.. SO EVERYTHING I SAID TODAY IS WRONG.
- "REG BLUE" (SPECTROSCOPE) WE SEE Y-G-B-I
- DARK BLUE (") WE SEE I-V

Y + "REG" B = ROY - G - B-I
Y + DARK B = RO-Y - G - B - I - V = FULL SPECTRUM ... SO WHY DOES Y + DARK B GIVE US BLACK ON THE OVERHEAD & Y + REG B GIVE US GREEN?

R + G = R - G = • BLACK ON OVERHEAD
• DARK GREENISH BROWNISH COLOR ON OVERHEADS

OTHER GROUPS

① LIGHT — RED FILTER — ONLY RED ALLOWED THROUGH
LIGHT — GREEN FILTER — LIGHT
LIGHT — RED FILTER — GREEN FILTER — EVERYTHING IS FILTERED GIVES US BLACK

* DARK BLUE MIGHT BE CONSIDERED PURPLE OR INDIGO

WHEN USING PAINT, MIXING RED PAINT & GREEN PAINT MAKES BLACK. THEREFORE PAINT AND LIGHT ARE SIMILAR AND REACT MUCH IN THE SAME WAY.

② BLUE
YELLOW
= ROY G BIV "WHITE"

WHITE
R, Y, O, G — G, B, I, V
YELLOW — BLUE

YELLOW LIGHT — BLUE LIGHT
YELLOW — BLUE

notebooks as a way to trace and document the thinking of the class or small group. In Emma's example (Figure 1.8), she and her group are trying to figure out what it means for one sound to be louder than another. This day, we gave students rubber bands to strum and asked them to describe the motion of the rubber band and the motion of the air around the rubber band. Emma's group noticed that the volume decreased over time and asked whether a louder sound caused air particles to move faster—to her group, a louder sound would both move more air molecules and give those molecules more energy, causing them to move faster and farther. As you see in Figure 1.8, Emma disagrees.

Emma's group thinks that a louder sound is due to faster movement, while Emma thinks the louder sound is due to more particles moving, although not moving any faster. We see this example as an important moment in the function of the notebook: Emma is not simply "taking notes" from the day's class, but using the notebook to explain and evaluate competing theories, much like a scientist would do. The notebook creates a space for a competing idea to be heard, particularly when a soft-spoken student is

Figure 1.8. Emma's Class Notebook

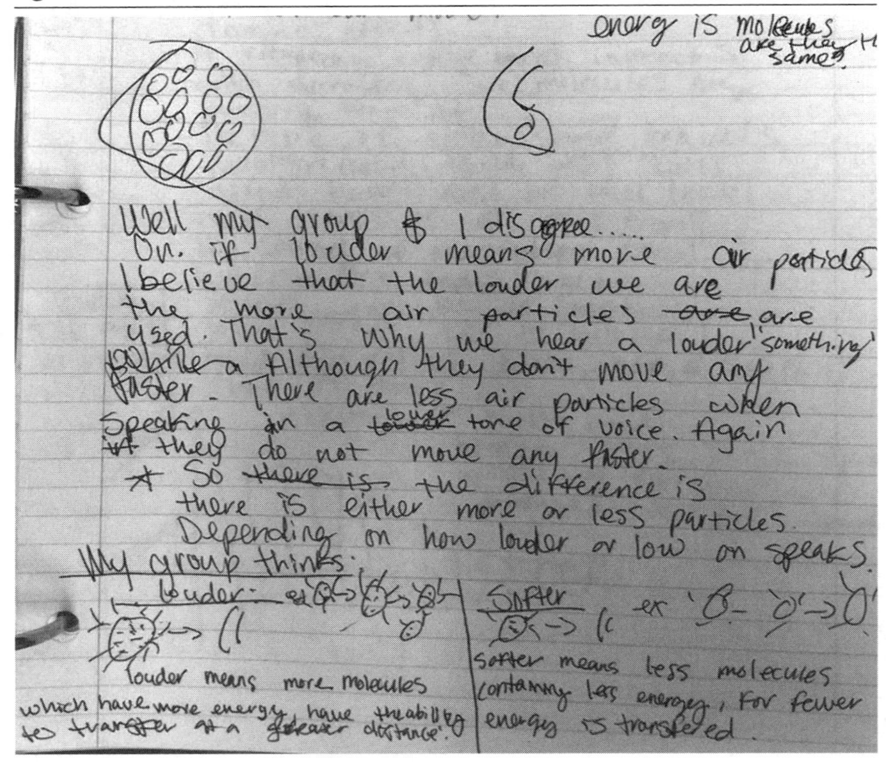

grouped with a strongly opinionated and vocal team of students, as was the case with Emma. The notebook functions as a platform for competing voices even when a student's actual voice in class is silent. Further, when a nascent idea in a notebook is validated by a peer, or by instructor feedback, the idea can be shared more easily with the class: The student has had a chance to try the idea out in a notebook before voicing the idea in a larger, and sometimes more intimidating, whole-class setting.

Finally, it is worth noting that Emma's question—whether louder is faster or louder is more particles—is not something that can be investigated experimentally given the tools available to the students. Any further progress on this question must be theoretical, and Emma's notebook is being used to sort out competing theoretical possibilities in the absence of additional data.

Kaitlyn

The final notebook example (Figure 1.9) from Kaitlyn offers a glimpse into how notebooks play into students' scientific lives outside of class. Like Linus

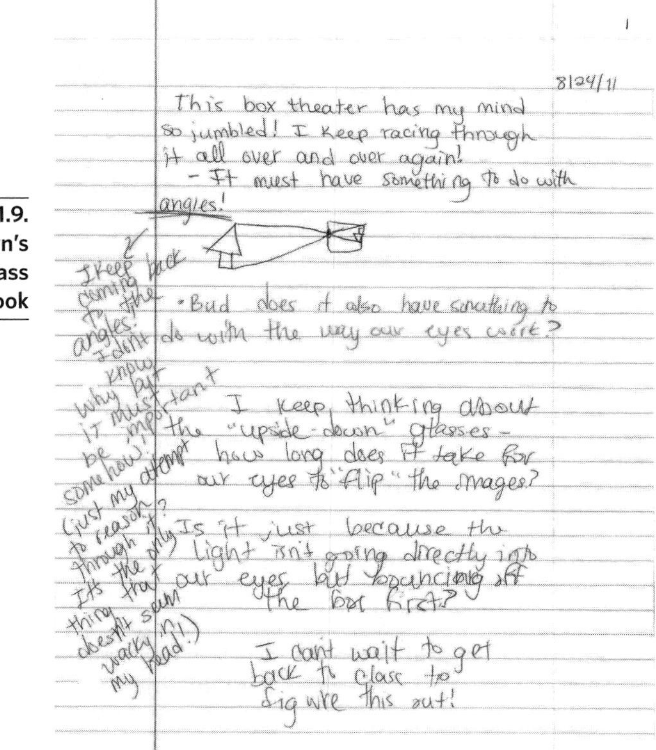

Figure 1.9. Kaitlyn's Class Notebook

Pauling ("Only last night in bed did I realize . . ."), Kaitlyn can't stop thinking about the science she's engaged in—the "box theater" (a pinhole camera of sorts) (Rathjen & Doherty, 2002). We find her at night scribbling down her ideas about how the box might work and declaring, "I can't wait to get back to class to figure this out!"

In many ways, this is the goal of a science course: We hope that the ideas and questions introduced during formal instruction are taken up by the students in their everyday lives, giving them a new way of seeing and being in the world. The lab notebook—particularly because of its unstructured format—supports some students in having and recording ideas outside of class time and class assignments. Ultimately, students take great pride in and ownership of their notebooks. In fact, one student, Janeal, sent us a note several years after taking the class: "Hi Leslie, It's Janeal. I just moved and was unpacking old boxes when I found my inquiry notebook. I am so proud of that thing! I just spent an hour explaining the human eye and pinhole theater to my boyfriend."

CHALLENGES

Time to Write

We have found in our own iterations of this course that quieter students who do not participate much in group discussions often have incredibly rich notebooks, and thus are participating in the ongoing conversation through their writing. In contrast, those students who are very active verbal participants often lack any notes regarding the conversation. For instance, Jessica's notebook mentions how she loves the class and forgets to take notes during class conversations.

For this reason, we have found that pausing class discussions periodically to give students 5 minutes to put their thoughts down on paper can be helpful. In addition, for very talkative classes, the last few minutes of class can be reserved for silent writing so that students can put down their thoughts and organize their ideas for the next class.

Getting to "Why"

Over the past few years, it has become clear that keeping detailed records of experiments and results is not what students struggle with the most in writing a notebook. Most students do this quite well. Rather, it is difficult for students to capture the reasons these experiments are being done in the first place and to make sense of their data and connect those data with ideas and experiments that others are sharing.

Usually experiments and observations in science are "theory driven"—they help us rule out a model, offer additional support for a model, or constrain the possibilities. There is a clear reason, then, that these experiments are being done. For instance, one semester students conjectured that you could not see a beam of light from a flashlight aimed away from you because the light behaved like little bullets shot out of a gun and therefore could ricochet off walls and bounce off other light "particles" traveling in a different direction. This light-as-a-particle theory implied that two beams of light from two separate flashlights would strike each other and alter each other's path. We tested it and the two beams of light did not interfere with each other in any way.

In general, our students' notebooks described the two-flashlight experiment and noted the observation that the beams did not interfere with each other, but they missed the connection to modeling. At best, students noted, "I want to see if two beams of light will change each other's path," not, "I want to figure out if light is more like particles as Dee mentioned—where beams can bounce off of each other—or more like the 'energy' that Steven mentioned and just passes through another beam of light." After the observation, students tended to summarize their findings as: "The two beams of light continued through one another without changing their path at all," not: "Now it is clear that light doesn't act like bullets out of a gun. We need a new model. Perhaps . . . ." In writing about science, we've noticed a similar tendency: students getting so wrapped up in the experiment that they forget the bigger picture of why it matters for science.

To help students attend to the "why," whenever a critical moment arises and everyone understands the implications, we give students 5 minutes to write and prompt them to think carefully about why the experiment is important, not just about the design of the experiment. At the end of the experiment, we give students further time to write and reflect on what the results suggest. Modeling this type of reflection in whole-class discussions helps students incorporate it throughout all the science they do.

Digital Notebooks

In this digital age, many scientists have switched to a computerized notebook kept on a shared server, an iPad, laptop, or blog (Giles, 2012). There are now a wide variety of companies that offer digital lab notebook and data-archiving services for professional scientists. We continue to play with ways of representing notebooks in both analog and digital forms.

One semester we tried giving students the choice between a more traditional paper notebook and a digital notebook kept on a laptop or iPad. We set groups up with a Tumblr account to keep a group blog about their investigation. We set up a Facebook page for the class to connect through.

We loved the digital photos that students took of their experiments (these were particularly useful when studying color; describing results of color experiments is difficult). This data archive was incredibly useful during "lab meeting"–style discussions when one group wanted to share its observations with the others in the scientific community. It was fascinating to visually follow the progression of students' experiments and to use their investigations in further assignments. Using Facebook allowed a "backchannel" of discussion to take place that, we think, made us feel more of a community and got other voices into the conversation. Digital platforms encouraged the continuation of conversations outside of class.

However, digital notebooks came with downsides as well. Students weren't able to rifle through the paper pages of a physical notebook to glance at early ideas and data. Students were less inclined to doodle and sketch ideas. Moreover, the group dynamics changed. Rather than research groups engaging in debates over their experiments, students got lost in their laptops and stopped talking to one another at all for large swaths of time. Our experience is not unique: Recent literature from professional laboratories also explores the benefits and challenges of electronic and paper notebooks (Machina & Wild, 2013) and refrains from strong claims of support for a digital or a paper approach. Ultimately, we believe in working with students each semester to decide which platforms support their thinking and address our goals for the course.

FEEDBACK AND GRADING

Peer Feedback

In general, peer feedback on notebooks is very informal. We've often seen students refer to one another's notebooks, as they work in small groups, to (1) remind one another where they left off, (2) bring someone who was absent up to speed, (3) recall an idea they had together, or (4) compare data or diagrams. These informal comparisons allow individuals to see what others are writing down (or not writing down) and make adjustments to their own notebook (for instance, being more precise and detailed with data collection or including more diagrams).

We also see notebooks make appearances during whole-class discussions or group presentations. A student often will grab a notebook as a kind of talking point. For example, one student, Carlie, shared a diagram from her notebook. The class had been puzzling over why the spot of light from a Maglite flashlight changes as you adjust the casing around the bulb. After Carlie shared her drawings during a class discussion, another student asked her to pass them around the class so that everyone could take a closer look.

Subsequently, her diagram and way of thinking made appearances in other students' notebooks. (Chapter 3 expands upon this vignette to show how Carlie and others made sense of the curved mirror in the flashlight through their diagrams.)

Instructor Feedback

We generally collect notebooks three times in our 15-week semester. The first notebook check is intentionally situated after only 2 or 3 weeks of class so that students get feedback on their notebooks with enough time to make changes to their practice if there are any issues.

We write feedback on sticky notes so that we aren't writing on top of students' work. These sticky notes are filled with comments, ideas, and questions that are meant to engage students in a dialogue about their scientific practice. Sticky notes mark productive lines of inquiry, intriguing seed ideas that should be pursued further, areas where procedures or observations are unclear, great diagrams, and places where a little more reflection is warranted. We also use notes to support students through frustration and disappointment. Science is inherently frustrating at times; noting the productivity of a failed experiment and commiserating over a dead end are helpful. Through these sticky notes, we try to engage in a dialogue with students.

Grading

In grading notebooks, we use the rubrics that students developed at the start of the semester. These criteria are tied to scientific practice and understood by students to be a meaningful part of their work, rather than requirements set by an instructor. Students are asked to call attention to entries that address particular parts of the rubric. For example, if a requirement is that the notebook include descriptions of experimental setups, students should select a day in which they did this well. Not every entry will meet every element of the rubric, but they should be able to find examples for all the criteria. We use that to help us more quickly scan the notebooks and assign a grade to students' work.

Take-Home Messages

- Requiring a highly structured student notebook neither replicates the way in which notebooks are used in scientific discovery, nor allows the teacher to anticipate in advance the creative and idiosyncratic ways in which students will generate and trace ideas. When the teacher engages students in understanding the *role* that notebooks play in science (by providing students with images of scientists' notebooks and leading a discussion about their purpose), students can develop notebooks that fill that role without having to adhere to formulaic structures.
- Notebooks are a place where students have the freedom to try on ideas without being afraid of being judged if those ideas are wrong. Grading notebooks should honor that: Do not grade on whether a student has a correct idea; instead, grade students' ability to articulate their ideas and observations, and the ideas and observations of their peers.
- For many students, notebooks play a critical role in idea development. Ideas, observations, models, theories, and diagrams collide within a notebook, and students wrestle with them as they write to learn. This kind of informal writing is, in large part, what we mean by "writing" in the scientific disciplines.
- Writing in one's notebook constitutes a form of participation by allowing quiet students (like Emma) a safe place to disagree privately. It allows instructors a window into the minds of students who are thinking and learning but not speaking up.

Creating and Sharing a Whiteboard

> This has been one of my favorite courses. . . . I have never enjoyed a science class. I think the difference has been that instead of memorizing everything, we were discovering the answers ourselves and working together. —Kimberly

Scientists typically learn how to do science side by side with more senior colleagues. This apprenticeship happens as we learn particular experimental, computational, or theoretical methods from others in our lab—how to create an indium seal for a low-temperature vacuum, for example, or how to write a Perl script for computational work, or ways to interpret the crossing of two chemical potential lines. Just as significant is the apprenticeship into scientific reasoning as we learn how to plan experiments, examine data, challenge methods, and propose, defend, and refine models. This apprenticeship happens in conversations with colleagues and advisers—both formal and informal—as students, postdocs, and faculty discuss and share their work long before it is polished and complete. While such reasoning, troubleshooting, planning, and modeling happen at all times during scientific work, it is often during the lab meeting that the ideas are most clearly articulated and shared (Dunbar, 1999) and offer novices a place to observe and practice these ways of thinking.

While we rarely think of these face-to-face, informal interactions as rich in writing practices, it is, for most scientists, impossible to imagine these conversations taking place without a whiteboard, chalkboard, or similar space where plans, models, and data can be shared, discussed, and transformed. Moreover, it is hard to imagine the more formal scientific writing being produced without these more informal interactions. Sociologists and historians of science have commented on the prevalence of writing in informal interactions among scientists (Latour, 1990; Latour & Woolgar, 1986; Rooksby & Ikeya, 2012; Suchman, 1988). Many new science buildings implicitly acknowledge this, "wallpapering" the buildings with whiteboards and chalkboards to support such informal exchange. While notebooks serve as one critical medium for informal writing—where a scientist generates, reifies, and collects ideas, questions, and observations—the whiteboard is often where the group does this work, and it is a critical place in which a beginning scientist is apprenticed into scientific writing and thinking.

So central is the whiteboard to scientific activity that it is not uncommon for a job candidate to be asked to give an informal "chalk talk" in addition to more formal presentations. This "chalk talk" is, quite literally, an informal discussion of research ideas with faculty at a chalkboard. Unlike the formal research talk, this is not a lecture given with a series of slides, but a conversation among peers. This type of presentation is useful in that it allows faculty to interact with the candidate in an unscripted, informal conversation that is characteristic of day-to-day scientific activity. But note that the word *chalk* in describing this talk is not incidental: Unscripted, informal conversation among scientists often involves the use of chalk at a board.

When considering the many forms of writing that take place in a science classroom, it is these more informal ones that strike us as the most important and most overlooked. Asking students to mimic the highly stylized and idiosyncratic forms of a journal article or literature review is of little practical use for most students, particularly nonmajors. Moreover, scientific articles are developed over time through a process that begins with notes jotted down in margins, data transcribed into a notebook, and ideas constructed on whiteboards, or the back of the (proverbial) envelope, with colleagues. Supporting students in producing a final written product requires explicit attention to the earlier informal writing practices that characterize day-to-day scientific inquiry.

Emphasizing informal writing practices in a classroom, rather than the formal, final products, may be a stronger approach to engaging students in scientific inquiry and scientific writing. Are groups of students discussing ideas, pen in hand, gesturing at the board, erasing and revising throughout their discussions? If not, it's hard to imagine that much science is getting done, regardless of how much their term papers adopt the characteristics of scientific articles. If, however, you look into a classroom and see students clustered around a whiteboard, arguing over a representation as they move back and forth between empirical work, representations, and the model they are refining, then it is reasonable to assume that the substance of their work and the final products of their inquiry will be richly scientific.

This chapter, then, focuses on the use of whiteboards. Like Chapter 1 on notebooks, it emphasizes a particular low-technology medium that we use in our classes. The use of whiteboards affords collaboration, revision, feedback, and quick dissemination of ideas. It has been taken up in scientific settings for those reasons. The emphasis in this chapter is how to use whiteboards in classrooms and instruction to promote collaboration, revision, feedback, and dissemination around writing.

On a practical note, over the past several years of teaching our inquiry course, we have taught in classrooms lined in chalkboards or whiteboards, and had smaller whiteboards, cut from relatively inexpensive tile board, for lab groups to use. More recently, we have purchased rolling whiteboards;

while these are recommended, they are not essential. The handheld whiteboard, used by groups to work through ideas, has proved indispensable. We use these both informally, as a space where students can develop ideas as they work in and across lab groups, and formally, where they serve a function closer to an academic poster or presentation. Both are discussed below.

TAKEN UP

In a science lab, ideas often progress from scribbles in notebooks and on whiteboards, to slightly more formal presentations among colleagues at a lab meeting, then a presentation to the broader community at a conference, and finally to a peer-reviewed publication. In this section, we describe the two uses of whiteboards: at the earlier stage of "developing ideas," where students work collaboratively around a whiteboard to hash out a possible model or explanation; and at the stage of "sharing ideas," where the whiteboard is more akin to a poster, and groups share their work with others. As always, the divide is not always that tidy, as we often find students hashing out ideas that they thought were polished, or sharing preliminary ideas more widely for critique and feedback.

Developing Ideas

In most classrooms, the whiteboard "belongs" to the instructor and is a space that is populated with right answers instead of conjectures and half-baked ideas. In our undergraduate science classrooms, students work in groups of four, and each group has a whiteboard and markers available. Even so, our students generally do not start the semester by using the whiteboard to hash out ideas with their peers. Rather, they use it as a place to write down final answers. They are not used to "owning" the whiteboards, using them as a place to hash out ideas, or offering an assessment of information on others' boards. For this reason, the first use of whiteboards usually is driven by the instructor's request, and often peer critique also is prompted by the instructor.

This may happen, for example, as we circulate among lab groups and suggest that they work out their ideas on the whiteboard, which is placed in the middle of their table, and perhaps point out that various students have sketched out different models in their notebooks. We encourage the groups to use the whiteboard as they talk through their ideas and work toward consensus.

In Figure 2.1, an instructor sits at the table (bottom left corner) as students from two lab groups sketch ideas about the path of light rays through a lens to address the question of whether these rays all meet up

Figure 2.1. Student Groups Collaborate Around a Shared Whiteboard

at one spot on the retina. The small whiteboard sits in the center of the table, serving as a shared space to work through ideas. Amber (seated at left, but obscured in this photo) has diagrammed a model of three beams of laser light passing through the lens, the middle ray undeflected, and the three rays crossing one another at different points; this is an explanation for observations she has made using lasers and lenses. Dee (front, right) holds a marker. She has sketched her ideas out in the corner of the board, indicating that the rays should all fall in the same spot, an idea that makes sense to her but is inconsistent with Amber's evidence. Breanna (back, center) gestures as she shares her own ideas in reference to those on the board. Dee's diagram, in fact, is closer to one that a scientist would draw: Parallel rays falling on a lens should all cross at what is termed the "focal point" of the lens. (Amber's observations suggest that the laser beams were not, in fact, parallel—although the students do not yet realize this.) The instructor prompted the group to sketch out ideas on the whiteboard, but she rarely intervened during this discussion. As this work took place midway through the semester, students were familiar with the class expectations and routines, and little additional prompting from the instructor was necessary.

Over time, the instructor rarely needs to prompt students to use the whiteboard, as they have become familiar with the routines and spontaneously reach for a whiteboard to diagram ideas or engage in a debate. It is these informal, unsolicited interactions that strike us as among the most reminiscent of our own scientific work. In Figure 2.2, Amanda (left) and Breanna (right) are working through ideas on how light rays must travel through a pupil and lens before coming into focus on the retina, and they are explaining their diagram to the instructor (middle).

Sharing Ideas

When it seems that most (if not all) groups are at a natural stopping point during a discussion or experiment, and have interesting data, ideas, or questions, we ask them to share those with the whole class using their whiteboard. In constructing those whiteboards, their work is closer to that of the lab group preparing for a conference poster session than to the more informal interactions as groups seek to make sense of early data. As they work to generate these boards, they often must reconcile differences within their group, determining what it is they want to share, selecting a representation that best captures their reasoning, and depicting the essential information on the board.

In one case, students had been working with a spectroscope and colored filters, and they noticed a pattern in the spectrum when white light was projected through overlapping colored filters. They had a fairly complicated method for determining the resulting color; working to articulate this on their whiteboard was challenging, and the students discussed exactly what steps they took so that they could describe their methods to the class. As in formal scientific practice, developing a presentation—for a conference, a lab meeting, or a more formal paper—forced these students to carefully articulate their methods and revisit their data to ensure their description was correct. Later, when they presented their ideas, their classmates found the description confusing, and one member turned to the whiteboard to demonstrate graphically—a far easier means of describing their methods. It is through these interactions—first with their small group and then with the whole class—that their methods were refined, articulated, and then re-presented.

Figure 2.2. Amanda and Breanna Discuss How Light Rays Must Travel Through a Pupil and Lens Before Coming into Focus on the Retina

As one student reflected after sharing a representation for star trails (long-exposure images of stars), "I figured out why stars don't make a full star trail, full 360-degree star trail, in exactly 24 hours using a diagram, but then when I tried to use that same diagram to explain it to others, it was very difficult for them to understand."

As with the informal use of whiteboards, this more formal use becomes routine as well. Students working in their small groups may come to a finding they want to share and may work to construct a whiteboard in anticipation of a whole-class discussion. In Figure 2.3, Amy shares a description of why our eyes must "work hard" to see objects that are close: They must refract the light more. The group shows one point source of light, a "close" eye, and a "far" eye. The closer one (shown at left) needs to bend rays sharply so that they reconvene at the retina; the farther eye (shown at right) intercepts fewer rays, and bends those rays less. The group had developed this idea on their whiteboard before revising it to share with the class. Developing, refining, and sharing of ideas become a seamless part of their inquiry.

Figure 2.3. Amy Shares a Whiteboard Describing Why Eyes Must "Work Hard" to Focus on Closer Objects

CHALLENGES

Encouraging Use

Some semesters, simply prompting students to use the whiteboards is enough: Students quickly use them as spaces to develop and refine ideas, first in their small groups and then between groups. Other semesters, however, students are initially less interactive—they report their methods or data and then look to the instructor for an evaluation. The whiteboards become a visual accessory to a presentation assigned by the instructor, not a place to work out and communicate ideas. For students accustomed to short, 1-day laboratory exercises in which they complete an investigation, report their findings, and receive a grade, a more extended inquiry and opportunity to revise ideas over time are foreign. Students are reluctant to challenge one another or suggest modifications to these "reports."

We have several strategies to help promote the use of whiteboards. As noted above, as instructors we encourage students to sketch out ideas on the whiteboard as we walk around the room. We might point to a diagram in a notebook and ask the student to sketch it out for all to see and work through. When we start the term with an initial prompt or question, we ask students to use whiteboards from the start, in their first predictions as a group. In this way, we show that the whiteboards are clearly a space for sharing preliminary ideas. For example, some semesters we start with an eye dissection, and we ask groups to first predict what they will see as they cut into an eye, and to write those ideas on the whiteboard.

Encouraging Feedback

In these early conversations, when students simply smile and nod at one another's work, we might select several boards, prop them up at the front of the room side by side, and ask students to compare them. Do the groups agree? How do one group's data inform another group's findings? If models differ, are they consistent with one another? If not, how can we come to consensus? In Figure 2.4, groups have been working on understanding the pattern of light from a flashlight. They recognize that the reflection off the back reflector matters and are trying to model those rays. Several whiteboards have been placed at the front of the room, and the instructor calls attention to the differences between the models before asking the students to discuss them.

Often the ensuing conversations are rich, as students describe the pros and cons of the various models and representations. However, we have had classes that are still reluctant to challenge one another's ideas, and the feedback they offer is often along the lines of, "I can see what you're doing. I just did something different. But yours is great, too." In these instances, we

Figure 2.4. Leslie Places Whiteboards Side by Side to Draw Out Critical Differences in the Representations

have modeled for students the kinds of interactions we expect, by leading a "mock" lab meeting, where the co-instructors, beginning with a lab group's whiteboard, discuss and refine the students' ideas, challenging and refining their board as we argue back and forth.

In Figure 2.5, the two instructors present students' data and talk through ideas together. Beginning with the original data (the spectrum of colors that project through different filters), we use students' ideas to predict the resulting color: a dark line added representing the spectrum that results, ideas about what color that might lead us to see, annotations of where the line represents a "blocking" theory of filters. By the end of the conversation, students have jumped in, presenting a model termed "multiplying fraction" that describes the amount of color that passes through a series of colored filters.

In cases when the class has had just one instructor, and modeling interactions in this way is not possible, we have led what we call a "master class," in which a student comes to the board to present a model and engage in a discussion with the instructor as the student works to refine the model. (These are based on master classes offered in music, where an expert will teach a student in a theater-like setting as other students watch, receiving a lesson by proxy.) We choose a student who will be comfortable defending ideas that are subjected to instructor criticism and who may have an interesting idea that we find worth pursuing.

One semester this took place after students completed a homework assignment on the eye that asked what an eye might see if it had no lens or cornea, but just a pupil in front of the retina. The assignment was designed

Figure 2.5. Leslie and Irene Lead a Mock Lab Meeting to Refine a Whiteboard

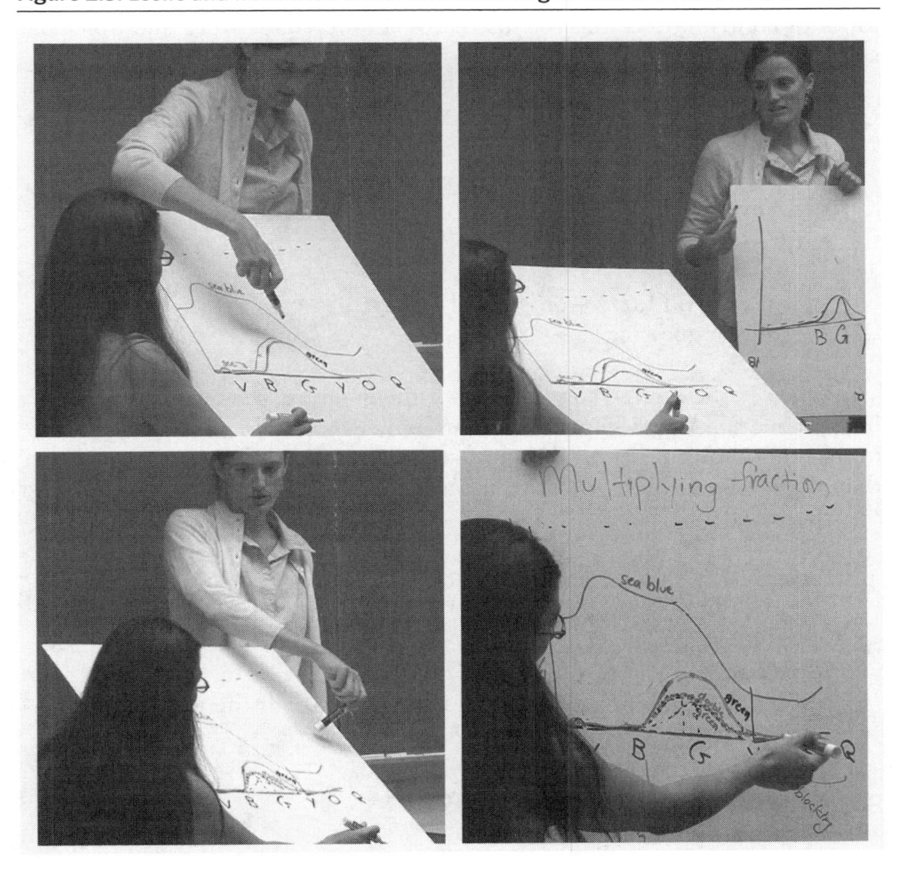

to bring out the range of ideas that students had about what optical problem a lens must solve in order to see clearly, to promote precise models of what it means for an image to be blurry, and to draw attention to scale (the eye is notably smaller than the pinhole theaters we have used, which impacts the blurriness). A student named Daniel was asked to recreate his homework assignment at the front of the room as the instructor asked a range of questions regarding the rays he was drawing: why he chose to draw those rays, why they were drawn in that way, what implications his diagram had for the image that would be seen, and whether it was possible to construct a quick experiment to verify the predictions. (All students in this class had done the same homework assignment, so the ideas that Daniel was constructing could be compared with their own work.) In the transcript below, Daniel is drawing a model for an eye looking at a stoplight; references to the "light" or "bulb" are in regard to the bulb of the traffic light.

As Daniel worked at the board, drawing and redrawing rays, the majority of contributions that the instructor (Leslie) made were questions and suggestions such as:

- As you're drawing, tell us why you're drawing what you're drawing . . .
- So tell us more—why, why those three?
- Okay, why did you choose the one that you chose?
- So talk to me about those . . .
- Why blurry?
- Is it red and fuzzy—or is it red and crisp?
- Are you guessing?

When Leslie offered suggestions for the diagram, they were ways of being systematic. As it became clear that Daniel simply drew rays from a top, middle, and bottom region (but not, say, the topmost ray or some other ray that might help define a boundary), Leslie made a suggestion to focus on just one spot on the bulb:

Leslie: . . . let's focus on the very top of the very top bulb. And there's lots of light leaving that spot. Like, I'm not worried about all the light, right?
Daniel: Yeah.
Leslie: But I am worried about all the light that left there that got into the eyeball.
Daniel: Okay—I can do that.
Leslie: So talk to me about those rays.

From this spot, Daniel drew three rays that entered the pupil: the topmost, bottommost, and a middle ray. He described it as creating a spot of light on the back of the eye:

Leslie: What does that spot look like? If you were inside the eyeball looking at the back of the eyeball, . . .
Daniel: Like a big, blurry, red splosh.
Leslie: Why blurry?

Daniel clarified that, overall, the image was blurry but this one point on the traffic light would generate a crisp red image of a circle in the eye. At this point, Leslie suggested the class should model this with a Maglite and aperture, seeing, indeed, a spot of light. Daniel continued with his model, arguing that each spot on the traffic light creates another spot of light, so that there are what he called "overlapping Venn diagrams of color."

This master class lasted most of an hour. While Daniel worked, Leslie was modeling how peers might ask questions, critique, and improve one another's work as they shared their ideas on the whiteboard. She also was highlighting ways in which students can be more careful and systematic in constructing their whiteboards. And finally, she moved between predictions and experiment; as Daniel considered one point of light, and then two, she used materials to confirm his predictions. By the end of the hour, other students had entered into the conversation, peppering Daniel with questions of their own as they sought to understand and build on his ideas:

> "So are you saying the green and the red mix in the middle (to make yellow) and that would add to the already yellow light?"
>
> "The Venn diagram circles—those are each one piece of the stoplight?"
>
> "So each [Venn diagram] circle is one ray?"

One student compared the small pinhole to the large pupil by suggesting the pupil lets through "spaghetti" and the pinhole "angel hair." Daniel corrected this, saying that the rays "don't get larger, they get separated." And Kaitlyn followed up, asking, "So that's why they would be blurry?"

Encouraging Multiple Voices

Finally, while not common, there have been times when our courses are populated by a few extremely talkative students who engage in extended dialogue with one another, while others have trouble getting a word in during the whole-class conversations. We will discuss some methods for engaging more voices ("silent science" and "gallery walks") in Chapter 4.

Finding Productive Questions

Not least among the challenges is one of instructional design. If you find your students are struggling to have productive conversations around whiteboards, or to design whiteboards with meaningful content, the problem might be with what they are using those whiteboards for. Do they have competing ideas for a phenomenon that they are interested in understanding? Or do they have interesting data that can help further develop a model they are interested in refining? Are there various possible representations, and students are particularly proud of one they developed? If not, it's going to be hard to sustain a productive scientific conversation, regardless of students' knowledge and abilities.

One of the greatest challenges for our class has been to find questions that launch students into rich scientific inquiries and to shape the conversations to draw out those competing ideas. Stilted conversations and students asking, "What should I put on this board?" are often an indication that the instructor has not started with a meaningful question—not that students need more support in having conversations and using these boards.

FEEDBACK AND GRADING

Unlike a notebook, which is most frequently for personal use, or a paper, which typically is designed to disseminate fairly polished findings, whiteboards are meant to be shared, debated, revised, and refined. That is, they are designed for receiving feedback. Much of that feedback should come from peers—the lab groups as they construct the board, and the class as a whole as students share their work with one another. As noted above, the instructor may need to model the kind of feedback that students should offer one another. For example, the teacher might attend to the substance of the ideas represented on the board and the use of the board to further those ideas, rather than focusing on more superficial aspects.

In later chapters we offer some specific ideas for classroom strategies that instructors can use to structure peer feedback on whiteboards. For instance, homework, gallery walks, and silent science are methods we use to promote and improve the kind of feedback that students offer one another, and this carries over to their whiteboards.

We have experimented with grading "participation" in class, asking students to first do a self-analysis of their participation and then commenting on that. We have never been satisfied with this grading scheme, as students often comment on simply how often they spoke up. Instead, we now ask students to develop portfolios of their work and include images (taken with their phones) or notes from whiteboards. One benefit to this approach is that it calls students' attention to the array of scientific writing practices that they have used, both formal and informal, and to reflecting on how they have improved in the use of those practices. We ask students to include a cover letter—a reflection—with the following guidelines:

> We want to know what kind of writing was and wasn't useful for you and why. Please write a brief reflection that addresses these questions:
>
> 1. How useful was the peer and instructor feedback on your drafts?
> 2. How did you make use of the feedback in your revision work?
> 3. What kind of writing was the most useful to you, in terms of thinking about science, and why: the notebook, the whiteboard work in groups, silent science, homework, take-home exams?

We find that this reflection offers students a chance to think about how their ideas have grown and developed over time and the structures that support that growth. The portfolio also honors idea development as opposed to nascent ideas.

Take-Home Messages

- Whiteboards play an indispensable role in scientific research as a place to develop ideas with colleagues and to make that development visible as newcomers are apprenticed into the lab. Whiteboards can serve a similar role in the inquiry classroom: a semipublic, shared space where students work together to construct, share, and refine ideas, and witness how ideas are developed.
- This informal writing often is overlooked as a form of scientific writing, but it is exactly the kind of writing that "writing across the discipline" programs seek to encourage: specific and central to the discipline, and used in a way in which practitioners themselves "write to learn." Moreover, such writing cannot be taught in a generic "freshman comp" course, but instead should be featured in science content classes as students develop scientific ideas.
- Students are accustomed to the whiteboard as "belonging" to the teacher and as a space for right answers instead of drafts of ideas. They often will need support in using their whiteboards in a more informal and student-centered way to share tentative ideas that others are welcome to challenge.
- Instructors can model how to critique and co-construct ideas using a whiteboard. This may happen informally, as students present, or more formally through role-play in which the instructors position themselves as students working through a diagram, or through a "master class" in which an instructor and student critique and improve a diagram together as the rest of the class watches.
- In the classroom, whiteboards also can take the role of a conference poster. That is, they can be a place to share a group's consensus ideas with a larger scientific community for discussion and feedback.
- If your students still struggle to use the boards productively, reconsider the kinds of models they are engaged in producing: Students should be genuinely interested in the questions and phenomena they are working with, and should have ownership over the direction that this inquiry will take.
- Whiteboards are used to receive formative assessment and feedback from peers and faculty. We rarely provide written, summative feedback on posters; in-class interactions with peers and faculty are generally sufficient. We have opted not to grade this form of writing, although students are encouraged to discuss it in their portfolios and it is essential that they incorporate the ideas from whiteboards into more formal work.

CHAPTER 3

Refining Diagrams

> The times where I felt the most confident in the material were the times where I was looking at my fifth attempt at a diagram and everything started to make sense. —Justine

Some argue that natural philosophy became science when diagrams, graphs, figures, illustrations, and other nontextual representations were introduced to represent the objects of study (Cunningham, 1988; Edgerton, 1985; Lunsford et al., 2007). Social scientists lump all nontextual representations together and call them "inscriptions." Many researchers, most memorably Latour and Woolgar (1986), have noted how the entire work of a scientific research laboratory revolves around generating, manipulating, discussing, and crafting ideas around inscriptions: "Even insecure bureaucrats and compulsive novelists are less obsessed by inscriptions than scientists" (pp. 245–246). Thus, this chapter turns its attention to inscriptions and some of the practices we have employed to encourage students to generate inscriptions and to interpret and critique the inscriptions of others.

Many types of inscriptions are used in science. Far and away the most common of all inscriptions in scientific journals is the graph—on average, any given scientific research article contains approximately 3.5 graphs and two other nongraph inscriptions (Arsenault, Smith, & Beauchamp, 2006). Latour (1990) found that when scientists were unable to access their graphs, they "hesitated, stuttered, and talked nonsense" (p. 22) and were able to resume the conversation only when a graph was scribbled onto whatever scrap of paper was at hand. A primary reason scientists are so obsessive about graphs is that graphs are extremely persuasive (Latour, 1990; Roth & McGinn, 1998). Furthermore, graphs can reveal clear patterns amid the noise of real-world phenomena; can take ephemeral phenomena such as rat brain extracts and turn them into an enduring plot in a journal article; can transcend time and space; and may be combined, rescaled, annotated, and superimposed so as to highlight relationships among phenomena.

Although Latour focused his analysis on graphs, many other types of inscriptions share these same powers of persuasion (Arsenault et al., 2006). Conceptual diagrams (such as light ray diagrams, flowcharts, and electrical schematics) often are able to show the relationships between ideas or make the unseeable visible and thus convey an abstract idea to a wide audience. In some cases, diagrams are inextricable from the scientific idea they describe, as with Feynman's famous diagrams describing the behavior of subatomic particles. Other common types of inscriptions are photographs, illustrations, and videos that capture an image of the object under study (or the apparatus with which it is being studied). These have the advantage over graphs of being more concrete and closer to real-world experiences, thus recreating a scientist's experience for the reader. Still other types of inscriptions include tables, equations, geographic maps, chemical structures, DNA sequences, and even three-dimensional models (such as Watson and Crick's model of the structure of DNA).

Experienced generators and readers of inscriptions ultimately master the art of transforming inscriptions (Kozma, 2003; Pozzer-Ardenghi & Roth, 2010)—changing from inscription to inscription (as from a list to a table to a line graph or from a photograph to an illustration to a conceptual diagram) and transposing them (modifying an inscription through annotations, combination with others, rescaling, or other such alterations so as to produce new interpretations or highlight a point).

In this chapter, we focus on how diagrams and other inscriptions are constructed and improved in our course. Central to this work is an ongoing, iterative, and public work of constructing, sharing, discussing, and refining diagrams; connecting them with the parallel ongoing and iterative work from experiments; and adding increased precision, nuance, and abstraction over time. Through this process, students create rich diagrams, strongly tied to their scientific models, and with an understanding of the role of experimental evidence in the construction and validation of these diagrams.

TAKEN UP

In Chapter 1, we referred to Carlie's use of her notebook as she explained the dark and bright pattern of light generated by a Maglite (see Figure 3.1). Briefly, for those not familiar with Maglites, the reflector on the back of the flashlight is parabolic (a curved mirror), so that when the bulb is positioned at the focus of this parabola—what students eventually will come to consensus in calling the "middle" position—it will reflect a "beam" of parallel light rays. As the bulb moves closer to the reflector (the "in" position), the reflected rays glance off the mirror and fan out, leaving a darker spot in the center of the beam. In contrast, as the bulb moves away from the reflector (the "out" position), the reflected rays cross over the center line, again leaving a darker spot in the middle. Student diagrams below (Figures 3.2–3.4) gradually will come to illustrate this. This inquiry is one that emerged as students used Maglites to examine another question about light. Although we had not anticipated this line of inquiry, we pursued the question

Figure 3.1. The "Dark Spot"

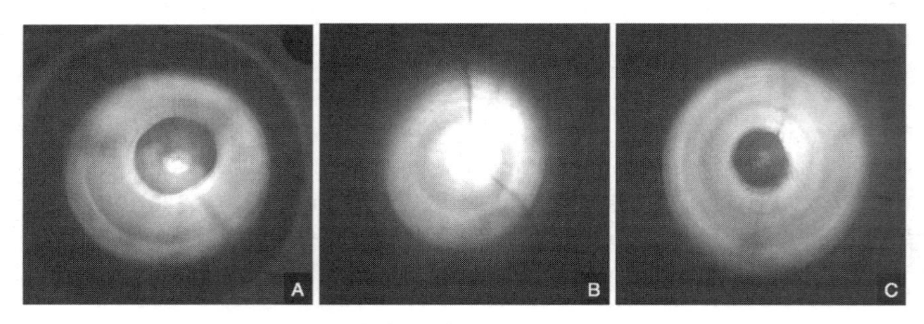

A: The Bulb Is Near the Reflector ("in");
B: The Bulb Is at the Focus ("middle"); C: The Bulb Is Far from the Reflector ("out")

found it generative: Not only were students able to create and refine models for reflection and images, but the techniques and insights that students produced were particularly relevant for our later investigation of lenses, which have properties similar to those of curved mirrors.

In this section, we present several vignettes from the weeks leading up to this shared understanding that highlight the ways in which Carlie and her peers use diagrams, how those practices change over time, and what classroom structures and instructor moves support these shifts, and we relate all of the above to the inscriptional practices of scientists. One of the things to notice throughout this section is just how contextualized inscriptions are—that is, their form is determined not only by the observations students are trying to explain but by the scientific questions students are wrestling with and the message the authors are trying to convey to their audience.

Week 2, Wednesday

By this point in the class, the fifth meeting of the semester, students had observed that in some bulb positions there is a dark spot, but in other positions there is not (see Figure 3.1). Moreover, if black tape is put over half of the curved mirror, then the shadow cast by the tape is on the same side as the tape in the "in" position, on the opposite side from the tape in the "out" position, and not visible at all in the "middle" position. The lab groups, however, were using different terminology for the bulb positions. Some called them "A, B, and C"; others called them "near, middle, and far"; and still others called them "in, middle, and out." To add to the confusion, the way the Maglite operates is that the bulb is "out" when you first turn it on and moves into the "in" position as you continue to twist the casing out.

Without our having clarified these differences in terminology, students were asked to create a diagram representing how they thought light rays from the bulb created the spots of light they saw. This day was an opportunity for groups to share their diagrams with the rest of the class and get feedback from one another and the instructors, similar in style to a lab meeting in a scientific research laboratory (see Chapter 2); through these conversations, the discrepancies in our use of terminology were brought out and clarified.

We started with Tina's group explaining its whiteboard (Figure 3.2). Part A of the diagram shows the bulb "in." The light bounces once off the mirror at an "acute" angle (more about what the group means by "acute" is described below) to cross to the other side and make a bright ring of light around the dark spot. Part B shows the bulb in the "middle" position and the light rays bouncing in order to ultimately converge in a small tight area in the center, thus avoiding a dark spot. Part C illustrates the bulb in the "out" position, with the light rays bouncing twice and crossing to once again create a bright ring around a dark spot.

As the group proceeded with the presentation, it became clear that the diagram was working backward from the observation to the theory. For

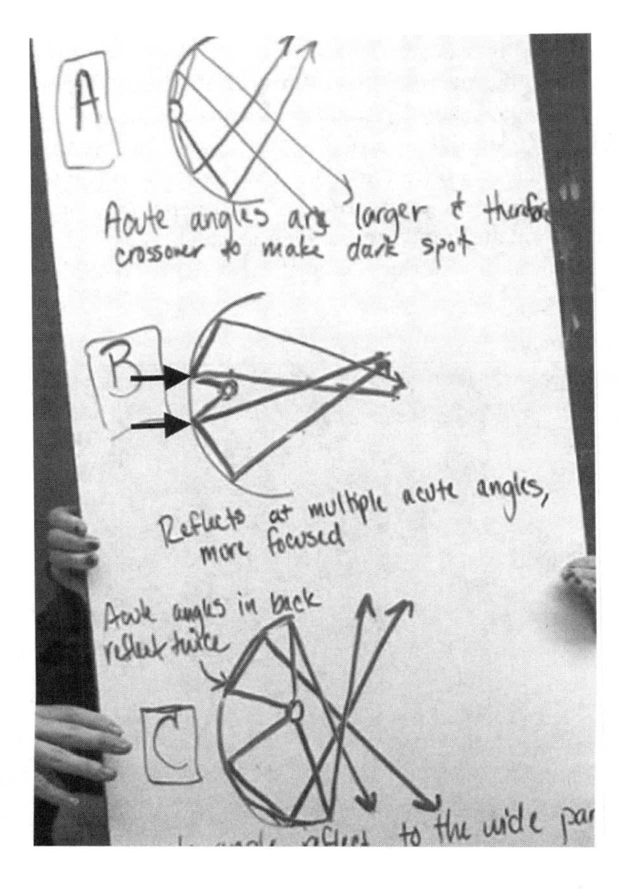

Figure 3.2. Tina's Group Whiteboard

example, when Irene asked why the light from the bulb ends up splitting in two directions at the points indicated by the black arrows in part B (not present in their original diagram), Alana suggested that "we know it's going to hit the center, so either it has to go straight out with one bounce, or it has to bounce twice and get to the same place, because we know it gets to a focusing center." They weren't sure whether it should be one bounce or two, so they drew both possibilities, guided by their conviction that in either case the light had to end up clustered in the center.

The students in Tina's group were tightly constrained by their observations. In addition, their diagram and presentation drew the audience's attention to the angle at which the light rays bounce off the mirror (a question they struggled with in their group) and the number of bounces (which they thought might explain differences in brightness). In the discussion that ensued, the instructors pushed the members of the group to more clearly define what angle they were referring to among the many on their whiteboard. Many groups had been wondering what happens when a light ray bounces off a mirror. Does it break into two rays or stay as one? Is there a rule to determine the angle at which the incoming ray will bounce off the mirror? Will it hit the mirror once or multiple times? Thus, it was very important that everyone in the class know exactly what one another meant when talking about "that angle."

The students in Alyssa's group began by saying, "I think we have basically done the same as [Tina's] group." This struck us as fascinating because their diagram (Figure 3.3) was clearly different from Tina's. Early in the semester, we tend to find that students are unaware of the differences between diagrams because novices focus on aesthetic surface features such as color and labels and don't know what actually matters scientifically (Kozma, 2003).

Figure 3.3. Alyssa's Group Whiteboard

Alyssa probably was noticing that both her group and Tina's group had rays that "cluster" in the center when the bulb is in the middle position and rays that reflect off the walls multiple times to "fan out" when the bulb is in the in or out position. To the instructors, however, the diagrams are markedly different: Alyssa's group drew the "bowl" of the curved mirror much deeper than Tina's group, thus prompting Alyssa's group to draw many more bounces within the mirror for all three bulb positions before the light ultimately escaped. The students in Alyssa's group had a more theory-driven approach (versus working backward from the observation as Tina's group did) in that they thought a bulb near the back of the mirror would have more time to bounce around in the mirror than a bulb farther out toward the opening. This idea drove their diagram, up until the point at which the light left the "bowl." Then, particularly with the middle-position diagram, you can see that the lines take a strange bend in order to get to the central area where the light is.

The "many bounces" idea of Alyssa's group attracted the attention of Marie, who noted that it can't possibly bounce multiple times, because when black tape covers half of the mirror, light still ends up on the screen. Marie comments on Alyssa's diagram, saying, "I think it only bounces once because that would explain the tape idea in my opinion. . . . Because black absorbs light, you are not going to get a reflection off of it. . . . So if it always bounces twice, then there would be no light shining out."

After a little discussion, the instructors brought Alyssa's group's diagram to the front of the class to annotate the drawing together. Irene added a black line representing black tape to one side of the curved mirror and erased any rays reflecting from the black tape. In this way, we modeled a theory-driven approach to diagramming—that is, one that works from a conceptual idea about how light should behave rather than working backward from an observation.

Just after this, Carlie introduced her group's diagram (Figure 3.4), one that she had been continuing to modify during class. Her diagram was

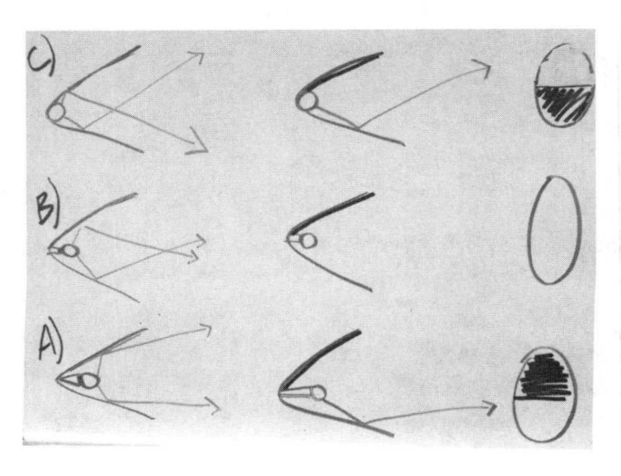

Figure 3.4. Carlie's Group Whiteboard

drawn in such a way as to account for the tape shadow as well as the dark spot observations. On the other hand, her diagram was not accurate. For instance, when the bulb is "in" and close to the reflector, the tape shadow should be on the same side as the tape, and when the bulb is "out" away from the reflector, the tape shadow should be on the opposite side from the tape. The ability to relate both the dark spot and tape shadow observations together on a single whiteboard was nonetheless a productive step forward.

Week 2, Thursday

The following day, we provided students with a template diagram of the Maglite, with a true parabola shape to the mirror and the bulb at its focus. Two groups ended up placing a flat mirror along the parabola on the paper and shining a laser along one of the ray lines around the bulb (see Figure 3.5) to see how the reflected ray would travel. As you can see, you get a pretty good approximation of parabolic reflectors—that any light ray from the focus that hits the parabola will reflect off with a light ray that is parallel to the line of symmetry. Other students proceeded theoretically rather than experimentally. For example, Sara used a protractor to draw tangent lines and angles of incidence, only to also discover that all the light rays exited the parabola roughly in parallel, corroborating the work of her group with lasers and mirrors.

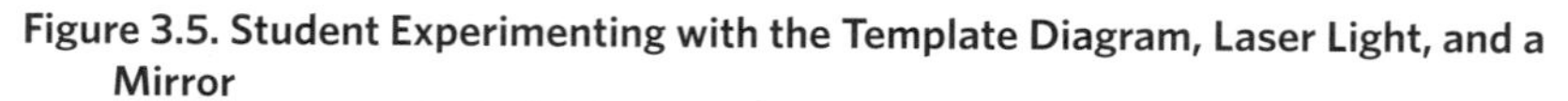

Figure 3.5. Student Experimenting with the Template Diagram, Laser Light, and a Mirror

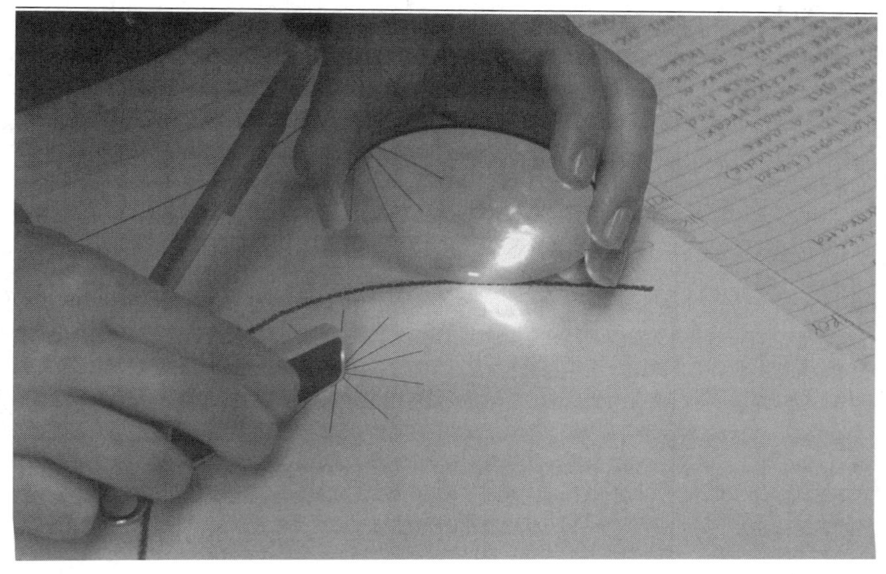

Figure 3.6. (A) Mary's and (B) Sara's Group Whiteboards

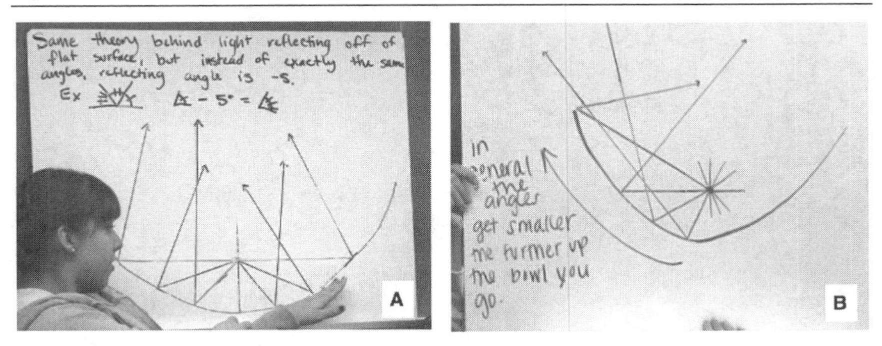

Week 3, Tuesday

The following class period, students were asked, as a group, to create a whiteboard describing their ideas. The six group diagrams differed in important ways, in part because of the different strategies adopted by the groups, but also because of what they were trying to communicate.

The first two groups to present (Figure 3.6) had attended to the angles at which the rays reflected off the mirror. They argued that instead of reflecting at "equal angles," as rays do from a flat mirror, the curved mirror must bend the rays more, and they offered some rough guidelines for how to account for that difference (e.g., 5 degrees, or "in general, the angles get smaller the farther up you go"). Both groups noted that these were ideas that seemed consistent with observations, but had little more to justify their claims. Sara's group knew that the bulb in the out position made the shadow from the tape appear on the opposite side, and then reasoned that rays that hit near the lip of the mirror must be more acute than those near the vertex. However, while the students were investigating bulbs in three different positions, they drew their ideas onto the template diagram with a bulb in one position, not realizing their mistake. We put both of these boards on display at the front of the room as a way to highlight the differences between models.

The students in the third group, Marie's (Figure 3.7A), were trying to communicate a different point; rather than drawing a single bulb with multiple rays leaving it in all directions, they drew three laser pointers aimed perpendicular to the line of symmetry so as to compare a single light ray from an "in," "middle," and "out" bulb. This representation is quite abstract. Instead of diagramming a real scenario, they focused on a particular ray from the three distinct bulb locations and followed just that one ray; it is left to the viewer to interpret the rest. Because they superposed the three conditions, it became easy for the viewer to compare what makes each

Figure 3.7. (A) Marie's and (B) Kaitlin's Group Whiteboards

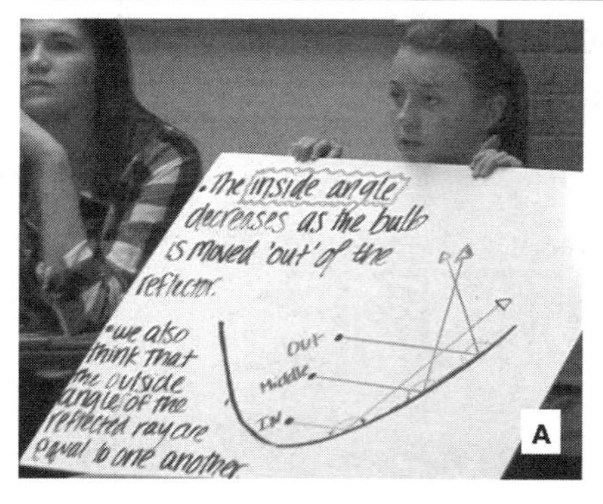

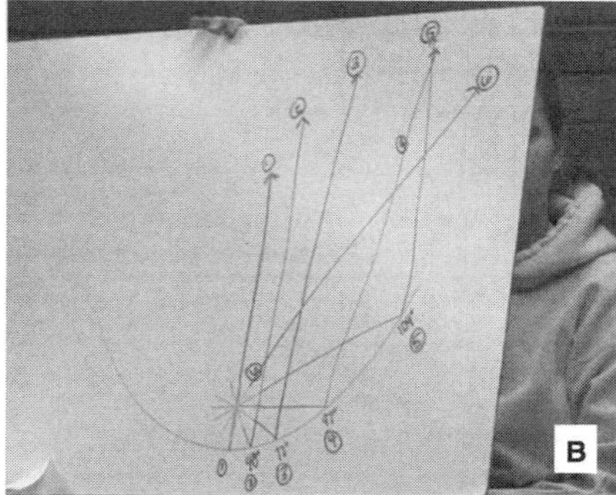

region unique: "in" are those places where the rays from the bulb reflect "away," "out" are those places where the rays from the bulb reflect "toward" the beam, and the "middle" reflects straight out.

The fourth group, Kaitlin's group (Figure 3.7B), used a laser pointer and flat mirror placed along the curve to experimentally determine the direction of the reflected ray. In addition, this group also considered rays that did not strike the mirror at all.

With two more diagrams ready to be put on display with the others, we asked the students whether both Kaitlin's and Marie's could be correct. This would, we hoped, call attention to the different ways in which students were referring to the "higher rays," something that was not clear as they described their findings:

> *Leslie:* I guess I'm wondering, could Marie's diagram and Kaitlin's diagram both be right? Because Marie's looks like as it moves up, the internal angle gets smaller, while in [Kaitlin's] it looks like as you move up, that internal angle gets bigger.

This question led to reconciliation of the superficial differences between Marie's and Kaitlin's diagrams. It also led to Marie's group recognizing that it had been experimenting with bulbs in three positions, but had illustrated the idea with only a single bulb position. In addition, Marie noticed that some diagrams are mutually exclusive, but not the two that, at first, had seemed to be in conflict, noting that her group's diagram showed a moving of the light bulb:

> *Marie:* But ours is moving of the light bulb. Theirs is moving—they are talking about one light bulb and many rays. We're talking about three different light bulbs. So those two diagrams [Marie's and Kaitlin's, Figure 3.7] I think are opposite. Versus that diagram of

ours and that diagram of [Kaitlin's] could both be correct. It's two different ideas.

Kendra (in Sara's group): We didn't take into consideration the three different levels. Ours got just where the light bulb was. I think that's where our confusion was. Our ideas were coming from different levels. While with the rays, the actual rays, there's a difference between it hitting [gestures with hand to show sloping down, horizontal, and sloping up] and how high the actual light bulb is. We were kind of thinking more how high the actual light bulb is, not where the actual rays are coming from. I think that's where the main confusion is. So you guys, yeah, your ideas correspond but I don't know if ours could. Does that make sense?

Leslie: So what I think you're noticing is that as the rays get higher, you get a smaller angle. But what you meant by the rays getting higher was the whole bulb getting higher. What they mean by the rays getting higher is that the rays just—the bulb just stays in the same place.

Kaitlin: So since all the rays are going straight up, it's possible to assume that this diagram we have is in the middle, not like far in or far out. Yeah. [Kaitlin places her board at the front of the room.]

Kendra now requests that Marie's board be moved up to the front, instead of her own group's:

Kendra: Can you put [Marie's] board up instead of that one [her own group's]?

Leslie: Do you like theirs better? Do you agree with [Marie's]?

Sara: Yeah, I like that one better.

Leslie: I think that one captures better what you meant by angle getting smaller.

Kendra: It does. Yeah.

The last two groups that presented (Figure 3.8) tied it all together. Both groups used a protractor against a tangent line at each point of intersection on the parabola. While Kimberly's group (Figure 3.8A) very precisely diagrammed just the middle bulb position, Marissa's group (Figure 3.8B) tackled all three bulb positions.

Kozma (2003) observed how chemists engage with multiple graphs, diagrams, and other representations of a chemical phenomenon and make sense of it. Through the interaction between the chemists, "what began as a disagreement turned into a shared understanding, as David and Tom together coordinated multiple representations to identify the product of their investigation" (p. 213). In the same way, our students were able to similarly find shared understanding about the underlying physics.

Figure 3.8. (A) Kimberly's and (B) Marissa's Group Whiteboards

Week 4, Tuesday Through Thursday

We conclude this section with Carlie. In contrast to the other groups and even her group mates who have been knee-deep in observations of dots on the mirror, chalk dust and lasers, and oddly shaped mirrors, Carlie has been focused entirely on painstakingly generating careful diagrams to show the "in, middle, and out" positions of the bulb. At one point she noticed problems with her sketches not replicating data. The instructors recommended that she reconfirm that the shape of the mirror and placement of the bulb accurately represented the real thing. She realized that the mirror in her diagram was the wrong shape, and went right back to the drawing board to try again. At another point, she had trouble figuring out the tangent, but worked through that.

Eventually, she generated very precise diagrams showing the in, middle, and out positions. After developing these ideas in her notebook, she shared this on a whiteboard (Figure 3.9). The class recognized that her diagrams presented an advance in quality, detail, and precision over what had been done before and set a new standard for the class. You can see this most clearly if you compare Carlie's final diagrams (Figure 3.9) with the initial diagrams from a week before (Figures 3.2 and 3.3). Not only are the ideas more scientifically accurate, but the diagrammatic techniques and conventions are more nuanced, accurate, and precise. The diagrams allow us to draw additional inferences—predicting places of bright and dark patterns we had not yet attended to.

CHALLENGES

Precision

There are different levels of precision for diagrams depending on the purpose the diagram serves —a quick sketch in a notebook, theory development in research groups, communication to others for a class presentation,

Figure 3.9. Carlie's Final Diagram

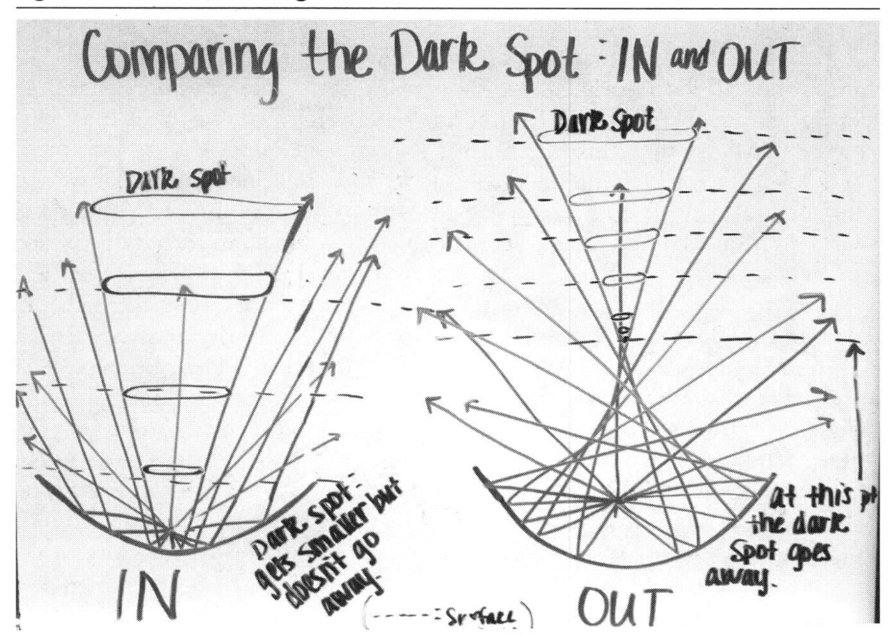

a formal writing piece, and so on. Fortunately, the level of precision called for will increase naturally and gradually as students grapple with an idea.

Sometimes, though, it is difficult for students to realize that there's a point at which precision in a diagram is more than an aesthetic choice, but becomes a way to make sense of and even generate new predictions that can be tested. For instance, just after Carlie presented her diagram, she said that there was still one thing that she did not fully understand and could not explain. Namely, one group had been drawing colored dots on the reflector and observing where images of these dots later showed up on the wall, relative to the dark spot. She wasn't sure whether her diagram was consistent with that group's observations. Thus, we went back as a whole class to connect the observation with the diagram that Carlie drew. In this case, the diagram and the observation corroborated each other and offered an opportunity for sense-making.

We have seen many other instances where a diagram and observation are inconsistent, thus pushing students to revise their diagrams and advance their thinking further. Even more exciting are the times when a diagram can predict something new that hasn't yet been observed, which we can then go out and try. Often these opportunities go unnoticed by the students so it is essential that instructors look for them and take advantage of them when they arise.

Labels

It's tempting to provide students with rules for graphs and diagrams—a rubric of sorts that will facilitate their work and make grading easier. For example, some instructors insist upon labels at all times. We agree that conventions such as these are important; they are included in style guides to every scientific journal and are a routine part of scientific writing. However, more critical for our course is to develop an understanding of the kinds of representational choices that students can make, and an appreciation for the role of context and audience in making those choices.

Figure 3.10A shows an example from Amanda's notebook in which she is trying to work out the geometry of how light reflects down a paper tube rolled around a flashlight. We choose to include this image because it is strikingly informal: no date, no annotations, no obvious questions being answered. It is the kind of page that might not be considered "good" note-taking in many contexts. And yet it reminded us of several scientists' notebooks in the way she is checking an idea on the back of a page. It's a side investigation; it was never discussed in class, but demonstrates that she is engaging with a puzzle and exploring ideas through her diagrams—deciding that

Figure 3.10. (A) Preliminary Sketch from Amanda's Notebook and (B) More Final Sketch

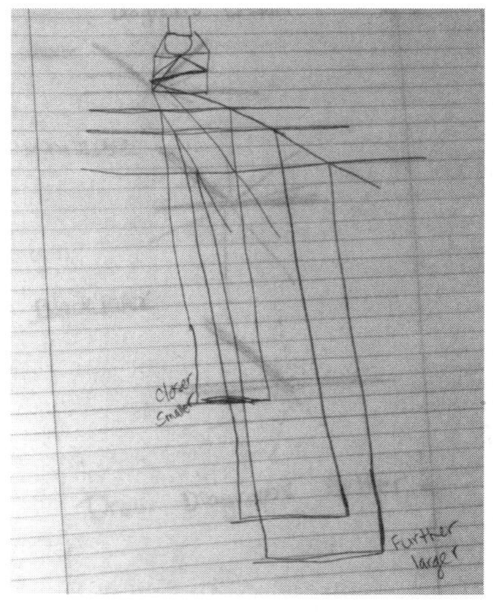

A

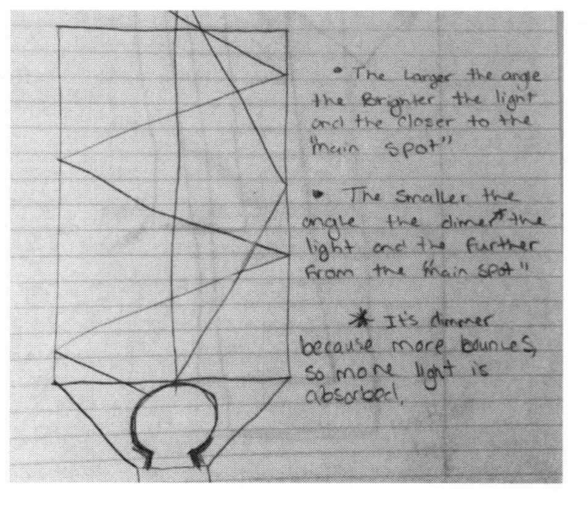

B

the steepness of the angle as it hits and disperses might matter, wondering whether distance might matter. Amanda's diagram does not require labels for the moment as her work is her own way of sorting through ideas on her way to something more developed that she might want to share with her group or the class.

In contrast, at times students' notes should be easier to interpret, with careful attention to labs and explanations that accompany them. In Figure 3.10B, we see Amanda's ideas reaching more of a conclusion, and she summarizes her findings in her notes. To be clear, this is still not a "final" diagram. We revise it yet again as we change our model for reflection from paper. Nonetheless, we can see Amanda tying together her predictions and sketches toward a more careful model.

Graphs

Several topics that we study encourage students to read, generate, and analyze graphs. Of particular note are studies of color where students work to interpret the Spectral Energy Distribution (S.E.D.) curves on the colored gel filters they are investigating (graphs showing percent transmission versus wavelength), studies of sound where students plot amplitude over time, and studies of motion where students record changes in position or speed over time. We usually end up spending a great deal of time helping students first to understand the graphs generated by others (e.g., those distributed by the company that makes the colored gel filters), then gradually to work toward generating their own graphs (e.g., by a spectrophotometer reading of a colored area), and finally to predict the results of future experiments graphically (e.g., given a red filter and a green filter with these S.E.D. curves, what curve would result if those filters were projected onto the same screen?).

2D Versus 3D Representations

As described in the NGSS, conceptual models take a huge number of forms. Diagrams and graphs may be the most common of these, but physical models also are used frequently throughout science. For example, one of the most famous of all physical models is Watson and Crick's model of DNA. While such models are not what first come to mind as "writing," the field of composition studies has pointed out that such representations are part of a broad definition of text, particularly multimodal text production, which often includes image, text, and sometimes even aural elements, as a way to make meaning (Kress, 2009; Shipka, 2011). For composition scholars, the decision rules we apply to composing any text should match audience and purpose.

We encourage our students to use three-dimensional models when it suits their purposes. For instance, one semester students spent a lot of time debating what happens to a light ray when it hits a nonreflective surface like a piece of paper. This led to the "half disco ball" theory, with the idea that the microscopic point on the piece of paper hit by the light ray acts like a half disco ball covered in tiny mirrors and sends a ray off in all directions. Daniel then went home and built a model of this with spaghetti and Styrofoam to illustrate the theory to his classmates.

FEEDBACK AND GRADING

Peer Feedback

There are many ways to offer peer feedback on diagrams. For instance, the idea of a master class, where other students eventually chime in, was discussed in Chapter 2. Several additional strategies, such as a gallery walk and silent science, are shared in Chapter 4.

There is another strategy that we use, which is very specific to diagrams. We briefly outline the idea here; it is described in more detail in an article published in *The Physics Teacher* (Atkins, 2012). The basic idea is to ask students to create a diagram along with a written description of the features in their diagram that they think are important (e.g., "The object is drawn the actual size," or, "Different colors were used to show the different light rays."). We then scan or photograph each diagram and post them on a free, online photo-sharing site like Flickr. In addition (and this is key), we generate a list of the features that students thought were important for their own diagrams and use their own words as a rubric to develop an online survey (through something like Survey Monkey or Google Forms) that asks students to evaluate one another's diagrams. In class, we discuss students' thoughts about which diagrams did the best job at each of the criteria and whether every criterion was necessary for every diagram (absolutely not—it all depends on the idea one is trying to communicate). Finally, students are asked to revise their initial diagrams based on the ideas generated from seeing others' diagrams and reflecting on the many ways ideas were communicated. After using this strategy for a couple of semesters, we are always impressed by the dramatic improvement in students' diagrams.

Instructor Feedback

In our first semester, we struggled to identify strategies to improve student diagrams. After a first assignment with incomplete and imprecise diagrams, we tried to tell students explicitly what to do and what not to do. Irene created a very long handout (six pages!) with "rules of thumb" and example diagrams. In her exuberance, she designed four diagrams—"not precise," "too much information," "incomplete," and "precise and accurate!"—and offered a detailed critique of each. This strategy failed. Students' diagrams improved very little, if at all, and we instructors ended up very frustrated. Now, after many more semesters under our belt, we realize that our attempt to give lots of explicit advice was actually an effort to make all student representations look exactly like the one that we would draw, when many equally valid alternatives were possible. The goal is not to make all student diagrams look like the instructor's, but to encourage students to adopt practices that help them express their ideas as clearly as possible.

Since diagrams are so contextualized to the idea and audience, instead of giving general advice, we make extensive use of questions and feedback on students' diagrams in context during class discussions (such as the master class), small-group research time (floating from group to group with advice as they work), and take-home assignments. Targeted, specific recommendations while students are drawing a diagram in class (such as, "Why not extend the lines from the object all the way to the screen?" or, "Would a protractor help here?") are often very well received. Too much advice at one time can be overwhelming, especially as students are simultaneously trying to develop ideas. As with science notebooks (see Chapter 1), we typically use sticky notes around a diagram to give detailed advice and feedback on written assignments so as to not scribble all over their original work.

Grading

When grading diagrams or graphs (that is, assigning a score), we tend to focus on the following two questions: How effective was this inscription at communicating the idea the student has in mind? and How well does the student's idea take into consideration the evidence and arguments that have been raised so far? Notice that we aren't judging how scientifically "right" the answer is.

Take-Home Messages

- Diagrams, graphs, and other conceptual models lie at the heart of science. Scientists rely on them to both generate explanations and communicate those ideas to others. They can serve a similar role in an inquiry classroom.
- Inscriptions are highly contextualized. That is, they are unique to the idea being communicated and the audience receiving the information. Providing a generic rubric for diagrams is insufficient for helping students learn how to make representational choices in their diagrams. Giving students authentic opportunities to generate, critique, and use diagrams in the context of inquiry, with a variety of audiences and scientific topics of study, is what we have found to be the most successful way to help students develop mastery over understanding and generating inscriptions.
- Comparing different students' diagrams can be a highly effective way to encourage students to consider a range of representational choices and better understand the affordances of those choices.
- Over the course of an investigation, diagrams will naturally become more precise, accurate, and complete as they move from informal, private musings in a notebook, to ideas shared with peers and instructors, to polished, careful pieces summarizing a large body of data.
- Students need help to become adept at generating and reading diagrams. We have found that giving students explicit, general advice is less effective than using contextualized feedback in the moment. Moreover, we recommend giving students frequent opportunities to give feedback to one another and to engage in whole-class discussions about best practices when communicating an idea through inscriptions.

CHAPTER 4

Peer Review of Ongoing Work

When I think about writing in this class, I think about revising. You are always revising, there is always something new to be learned and you have to add that to your paper. It's a never-ending cycle: You keep writing and writing and writing. —Ally

When most people imagine what it means to be a scientist or what it means to be a writer, they conjure a vision of the genius (often describing Einstein or Whitman) laboring away in isolation over a theory or typewriter. But those of us who are scientists or are writers know that the idea of the lone thinker could not be further from our lived experiences. Scientists share work—in pairs, in teams, in lab groups, in conferences, in poster sessions. We share work because we need the ideas of others to push our own ideas, to offer competing theories, or add insights we had not thought of. We know that feedback happens everywhere in both formal and informal exchanges: in hallways, in offices, in email, during lab group meetings, in Google Docs, in formal peer reviews, and even on social media when we pose a query to our communities. We also know that peer review functions within lower- and higher-stakes situations: with your colleagues and lab partners as you work through a nascent idea, at a conference where you get feedback on a more formalized idea, or very formally when you receive blind reviews as you work toward publication or to secure grant funding. As Harnad (1990) notes:

> The interplay of the prior ideas out of which [relatively noninteractive work is done] clearly consists of activities that profit from peer feedback. For most investigators the formal submission of a manuscript for peer review is not the first stage at which it has been subjected to peer scrutiny. That is what all those prior discussions and symposia and preprints had been intended to elicit. And all this prepublication interaction is clearly continuous with the lapidary stage at which the manuscript—usually further revised in response to peer review—is accepted and archived in print. Nor does it really end there, for of course the literature may respond to a contribution directly or indirectly for years to come, and there are even ways of soliciting post-publication feedback in the form of "open" peer commentary. (p. 342)

Regardless of the iterations, or the formality of the process, peer feedback is a necessary and invaluable part of the work of a scientist or a writer, and it is ongoing throughout all stages of an investigation.

As psychologists and social scientists examine the practices in research labs, the role of peer review—particularly the early and ongoing feedback that occurs within research labs—becomes even more clear. In a 1995 study, Dunbar noted that highly productive research labs had closely interacting groups with related but distinct research projects. Their interactions featured stronger group problem solving and more analogical reasoning as they shared their data and analyses. They were more likely to attend to surprising results when interacting with their peers and to use those results to inform significant developments; in particular, they were more likely to raise alternative hypotheses or generate new models. That is, the more a scientist solicits and engages in feedback with her peers, the more productive she is (also see Chapter 2).

Not only does peer review function to improve the author's ideas and the exposition of those ideas, it also plays a critical role in creating and sustaining a scientific community. This iterative, ongoing feedback is how we define what problems need solving, the challenges present in their solution, and the methods that will be generative in solving those problems. Moreover, it defines who is interested in the problems, who the intellectual stakeholders are, and what it will take to convince them of our claims. That is, we co-create an audience for our work via iterative, ongoing peer review.

Even though we engage in peer review all the time and see the value in our own work, it can be quite challenging to support peer review in our classes. Students are often uncomfortable taking on the role of critic or they devalue peer comments as not as important as instructor feedback. If we add to this mix faculty who have concerns about sharing the responsibility for feedback—Do students know enough to make productive comments? What if students' comments contradict faculty ideas? Or what if students' comments attend to the "wrong" kinds of things?—then it is no wonder that many people have less-than-ideal experiences with peer review in classroom settings.

The field of composition studies responds to these challenges by showing, through classroom and writing center research, why a focus on peer review is a valuable pursuit despite the challenges—offering examples of what is gained by students and faculty when we value peer review—and suggesting ways to support peer response practices. Jody Shipka (2014) speaks to the gains for both the writer and the responder, who have the opportunity to see multiple approaches to a writing task, when they share drafts:

> I believe strongly that students who are able to see (or in the case of the workshop sessions, hear about) the ideas, tools, techniques, and strategies their classmates employ while working on major projects benefit tremendously, both in

> terms of understanding that there is more than one way to accomplish a task, and in thinking about how the adoption of similar tools, techniques, or strategies might potentially impact their own work. (p. 226)

Shipka's argument parallels what we know about productive research labs: Seeing other "ideas, tools, and techniques" pushes us to do better work. Other composition scholars highlight the advantages of writing support that comes from peers because peers are responders, *not* graders, of writing: Peers stand in a different relationship with a writer and potentially can be seen "as someone to help them surmount the hurdles others have set up for them" (Harris, 1995, p. 28). Peers can function as a more friendly and compassionate audience, particularly as we work through difficult ideas. Ultimately, the key to recognizing the importance of peer review is an understanding of the social nature of learning and writing.

Literacy scholars have long argued that reading and writing are neither isolated practices nor a set of individual skills (Brandt, 1998; Gee, 1996; Street, 1984). Instead, literacies are *ways of using* reading and writing in particular contexts. The research of Scribner and Cole (1981) in the late 1970s and early 1980s clearly questioned the "cognitive" effect of literacy, and instead drew attention to literacy as a social practice:

> Instead of focusing exclusively on technology of a writing system and its reputed consequences . . . we approach literacy as a set of socially organized practices which make use of a symbol system and a technology for producing and disseminating it. Literacy is not simply knowing how to read and write a particular script but applying this knowledge for specific purposes in specific contexts of use. (p. 236)

New literacies studies continue to build on a view of writing as a social practice, particularly in a digital age as new technologies make the processes of writing and its use more visible. As Cathy Davidson (2010) argues in her work on 21st-century literacies, writing in the 21st century often occurs in networked, peer-supported, and collaborative environments. For example, by inviting anyone to collaborate in the production of knowledge, sites like Wikipedia disrupt the image of the lone writer or expert. And platforms such as Google Docs make the work of collaboration more accessible. We see value in the role of collaboration in knowledge production, and, in fact, see collaboration as a necessary element of meaning-making. But new literacies and ways of working require that we ask new questions: Whose ideas matter? What counts as expertise? How is knowledge distributed among people and things? And how do we make sense of competing ideas? We hope that our classrooms can challenge beliefs about where science happens and who can contribute to science. In the classes we teach, we support students as they try out various forms of feedback structures, reflect on feedback, and develop as authors and reviewers of ideas, all in a culture of caring. We see these practices as crucial in the 21st century, where people are called on daily to wade through a tidal wave of ideas and to comment, counter, and push on those ideas in ways that allow them to be heard.

Knowing that peer review is crucial to the work of scientists, we ask our students to participate in peer response practices through a range of structures throughout the course. In fact, as you will notice from other chapters, peer review weaves in a continual, recursive way in every activity we do. At times, the feedback happens through small- and large-group conversations, and at other times, feedback is offered in formal and informal writing. While we address approaches to peer response within the various chapters—explaining more specifically how peer response connects with particular activities—in this chapter we highlight two activities—silent science and gallery walks—intended to encourage and develop student identities as peer reviewers and colleagues. The purpose of this chapter is to focus on *ways of being* in relation to peer review: to show how a collegial stance toward idea development and feedback is modeled and supported in a scientific community.

TAKEN UP

In our class, peer feedback is not an activity mediated by red pens and final drafts. Peer review processes occur daily in an iterative process as students comment on one another's ideas—both orally and in written forms. Peer review activities give students an "audience" for their work: Other students present questions, challenges, and contradictions that the author can address in her writing and in clarification of scientific ideas. When we do turn to more formal approaches to peer feedback on final, high-stakes papers, we have already created a community of colleagues who care about their peers' ideas and who have response practices to draw from.

Silent Science

Silent science is an activity we borrow from English classes (where it is known as silent discussion). We use this structure to support student participation: We want to create a classroom space where everyone's ideas are heard, taken up by the class, and valued. The goal of silent science is to create a way for everyone to participate in giving and receiving feedback on ideas. The "discussion" happens on paper, which removes the fear of talking in class for some students and creates a structure where everyone's ideas are seen. The activity set-up is fairly simple: We usually begin by creating a prompt for students to write about or diagram. We create prompts that are familiar to students—an idea they have thought about before and that all groups can weigh in on, even if the class has not reached a consensus. Writing to the prompt takes students a half-page or less. We tend to choose an idea that was part of a homework assignment; this ensures everyone has a response that has received some consideration and allows students to pull out one particular part of their assignment for peer feedback. For example, when studying optics, and after a homework on lenses, we might ask students to explain what it means for an image to be "in focus." Or we might ask, when studying color, that students explain what makes a color a primary color. We also have found the silent science activity to be a good opportunity for students to receive feedback on diagrams; during a unit on pinhole cameras, we might ask them to draw a diagram of what happens when the pinhole is made bigger.

After completing a response in fairly short order, students sit in a circle and pass their paper to a peer next to them. The peer writes feedback, commenting on the ideas and the ways in which the diagram represents the ideas. The students ask each other questions and pose alternative explanations on the writer's paper. After a few minutes, the draft is passed along again to a new peer. This continues through multiple passes. In the end, the writer's draft is returned: She now has feedback from many peers with a range of feedback to support, challenge, and question the diagram and ideas. Often, we then ask students to attend to the feedback by revising their original diagrams and ideas or sharing the feedback with their lab groups.

In our unit on light, we begin by asking students to build pinhole theaters (Figure 4.1). Pinhole theaters (also known as box theaters or camera obscuras) are a simple way to create an image: As light passes through a small hole in a darkened box, an inverted image appears on the opposite

Figure 4.1. Students Exploring with Pinhole Theaters

wall. When students wear these boxes on their head, with the hole in the box placed behind them and ambient light carefully blocked, they can see that inverted image (Rathjen & Doherty, 2002). Using pinhole cameras to study light and color provides a rich experience for questioning how images are seen, which leads to models of how light interacts with objects, how light travels, and how color manifests in images.

In the two examples below, students were asked to take 5 minutes to diagram what they were seeing when they created pinhole theaters, or to write down questions they had about pinhole theaters. This occurred after the class had discussed and worked to create tentative models for their observations. The peer responders were asked to build on the writer's ideas or to ask questions about the diagram or explanation: What does this line mean? Why did you draw this here and not there? How does the whole image squeeze through the hole? In the examples below, we provide one example from a student whose diagram lacks a clear model, and where peers work to use her ideas to construct a representation consistent with them; and another example from a student who uses the diagram to explain a complicated observation that others don't quite understand.

Kayla's initial diagram is more of sketch of the set-up than a representation of a model. Her text offers some explanation, and her peers build this into her diagram, revising the diagram itself in order to map her words to the image and constructing the complete path taken by the light rays as they create the image Kayla has drawn (Figure 4.2). We see evidence that students value their peers' ideas and use a supportive, collegial tone in the feedback. For example, the use of a simple phrase such as "so I think you are saying," before adding a new diagram, shows that the responder is working with the writer's ideas, as opposed to correcting the writer or simply replacing her ideas with the responder's own. The responder takes up Kayla's description of "one line of sight" and shows how Kayla's written description could be better supported by more clearly showing the lines of sight in the diagram as well. The other responders work to clarify the role of light rays in "carrying" the image, asking a range of questions about how images are made through light.

Trevor's silent science example (Figure 4.3) offers another model. Students were noticing that images in the pinhole theater were inverted—both left–right and up–down—as compared with the "real world." But the students in Trevor's group noticed something strange: When they used the pinhole to look at writing (as opposed to objects), it was upside down, but did not seem to be inverted left–right. Trevor's group ultimately realized that this had to do with the perspective of the viewer (when you turn around to view the image behind you, what was on your left is now on your right). He attempts to explain this in his diagram by showing how the "view that would be facing the pin hole" started off inverted and then was "re-inverted" so that it looks like an "F" once more.

Figure 4.2. Kayla's Silent Science Draft

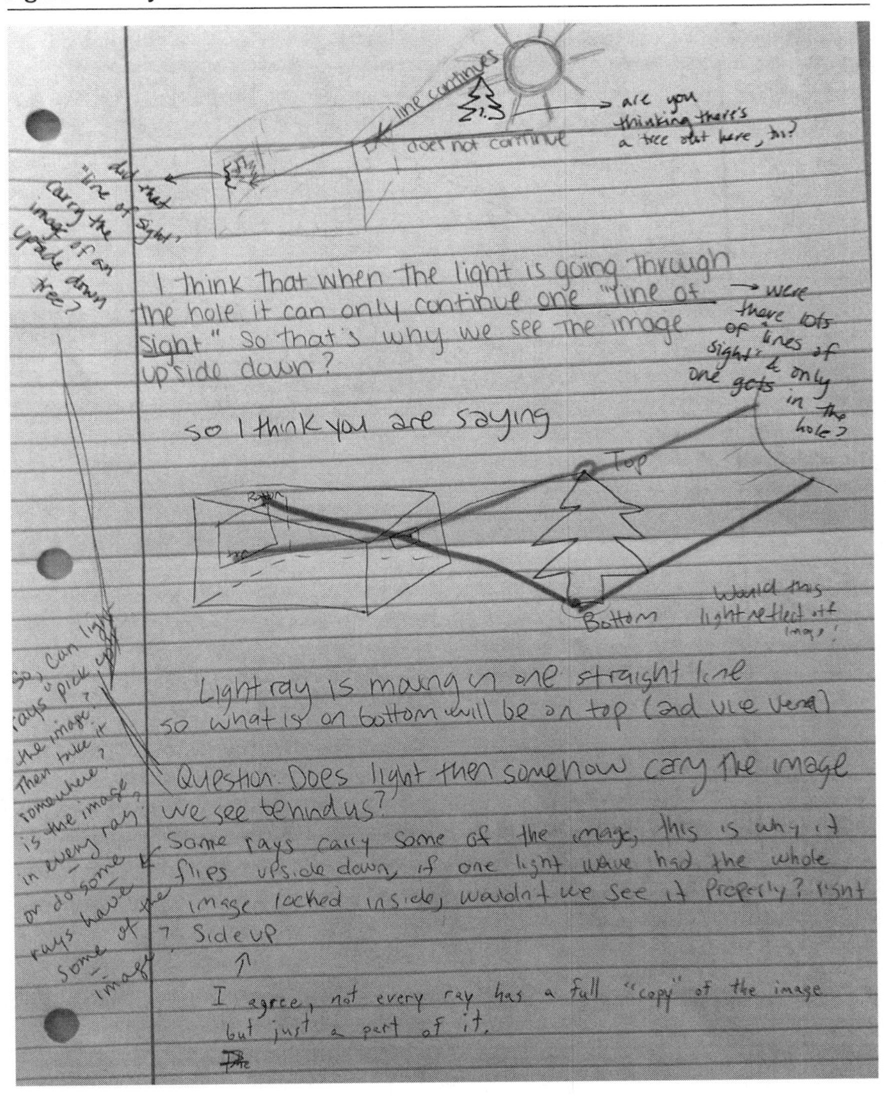

However, other students had not made the puzzling observation that Trevor worked so carefully to explain. Absent that, they did not understand why Trevor would draw an "F" as viewed from behind or clarify the view from the pinhole. Through the comments, Trevor might realize that he needed to help his readers understand the dilemma to be resolved and provide more context. That is, he needed to engage his audience in this question and why it was relevant to their own work.

Figure 4.3. Trevor's Silent Science Draft

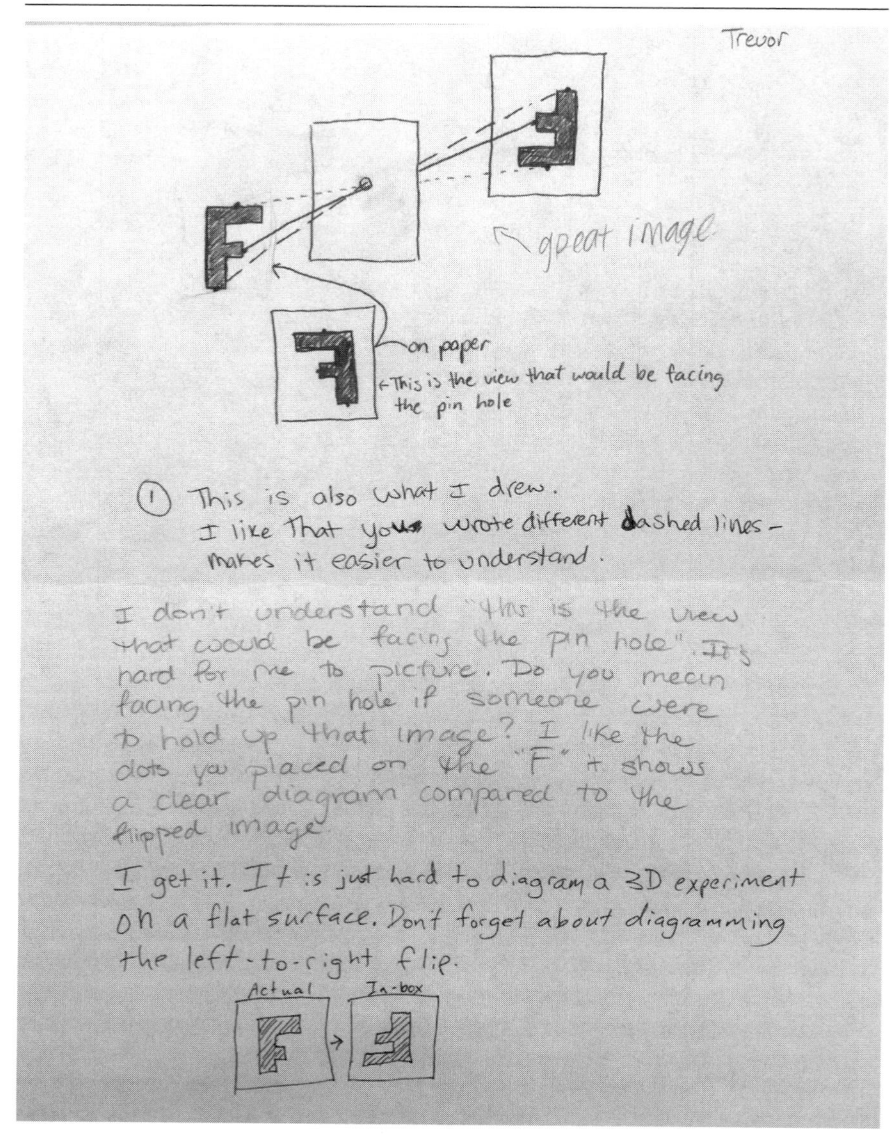

We see a range of advantages to incorporating an activity like silent science. Students see multiple representations and models of phenomena. In a class with very vocal and very quiet students, silent science provides a space where quiet students can be heard: Everyone's ideas are circulated and given serious consideration. Students come to see how understanding a scientific idea means not only knowing why "right" ideas are right, but why "wrong" ideas are wrong. The activity focuses attention on one, often brief, explanation—drawing students' attention to a particular claim, representation, question, or idea. This stands in contrast to feedback we often see on student writing, which tends to focus on idiosyncratic grammar rules and other editing suggestions, instead of attending to ideas. As a reviewer of ideas, a student must articulate this to peers just as we would do as a responder to an academic article. Finally, like many of our class structures, the silent science activity positions student peers as experts and authorities on scientific ideas. As we note in the chapter on final papers, all writing has an audience—and this course strongly positions the classroom community as the audience for our writing. Silent science is an important way in which we develop that audience.

Gallery Walk

Every few weeks, we ask students to participate in a gallery walk—an approach with many different manifestations in science education (see, for example, Francek, 2006). Like silent science, gallery walks are one of many activities that allow the class to see the various ideas represented in the room. In many ways, the gallery walk has characteristics of a poster session at a conference: Peers stop by, they ask questions about the work, and ideas are repeated over and over again with each new audience member. When all goes well, the presenter leaves a poster session with new ideas and more questions for thinking about the work. We tend to use the gallery walk structure when ideas are becoming more fully formed, as a way to work toward greater precision in our descriptions and inscriptions.

In our classroom, each lab group is responsible for sharing a representation of a scientific idea on a whiteboard. After the group has collaboratively created a diagram, we ask students to nominate one person from the group to be the spokesperson. This student stands by a movable whiteboard. As groups circulate around the room, the spokesperson explains her group's diagram to the other teams who stop by. In any class session, each group will receive feedback from three or four teams. More important, in the process of explaining the representation multiple times, the spokesperson often changes and hones the group's diagram and ideas with each retelling.

In the example that follows, our students were assigned the following homework: Diagram "how the light rays from a six-bulb flashlight would hit the retina in an eye that has an iris (with pupil) and a lens behind that pupil." In class the next day, the students in each lab group were asked to either (a) select the diagram that they thought best conveyed the group's ideas, (b) create a new diagram that was better than anyone's diagram at their table, or (c) if there were conflicting ideas at their table about what diagram was best, choose two diagrams that best illustrated the competing ideas. The groups then put their final selections on their standing white-board (Figure 4.4).

We focus here on Jonathan, who is the spokesperson for his group, as a way to show how ideas are developed through multiple retellings and feed-back from peers. His group has been investigating the shape of a lens and its effect on imaging a single bulb, so this question, which asks the students to consider multiple sources of light, is a challenging one. The group has created a diagram that shows three rays from each of two of the six bulbs. Each bulb is separately focused on the retina, with the lower bulb in focus near the top of the retina and the upper bulb focused near the bottom. Below on the whiteboard, the group has added information on how the shape of the lens affects whether the image is in focus, adding details most groups would not have considered. In particular, this group has sketched a flattened lens, with muscles pulled taut and the rays failing to converge; a rounded lens, with muscles loose, and the rays converging before reaching the retina; and a "just right" lens with rays converging at the retina.

As Jonathan explains his group's diagram, his peers—who have been working on different inquiries—ask questions about their own areas of interest. For example, one group has been curious about where the light rays bend, and argues that they "turn" as they enter and exit the lens. Noting that Jonathan's diagram does not show this, they question this choice. Another asks about the role of the muscles in changing the shape of the lens, asking whether, instead, the muscles might move the lens back and forth (not unlike the motion of a lens in a camera). And yet a third group is thrown off by how Jonathan describes the rays crossing one another: The members of

Figure 4.4. Jonathan Explaining His Group's Diagram, as Amy (from Another Group) Sketches Below

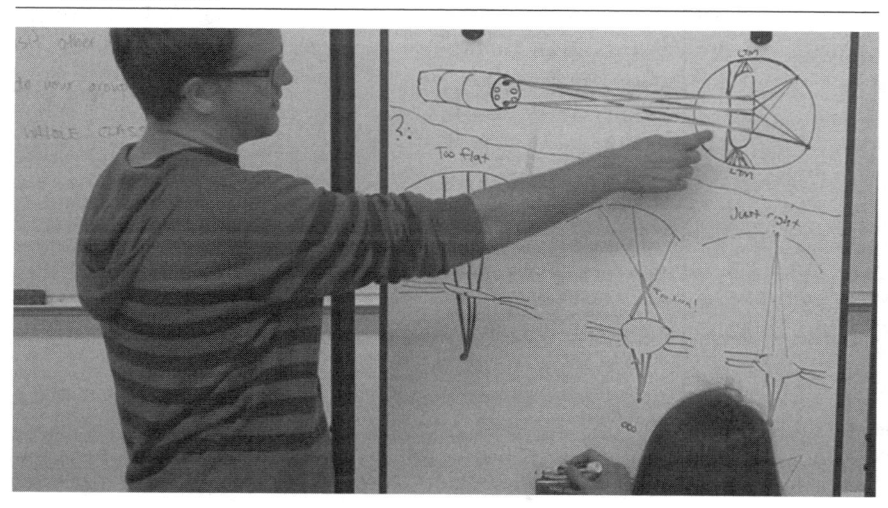

this group confuse the rays from separate bulbs crossing (which happens in multiple places) with the rays from a single bulb crossing (which happens at one place and is what creates focus).

Each of these ideas is taken up by Jonathan in his repeated explanations: With regard to the lens, he notes to later groups, "Something happens in the lens; we're not sure exactly what," as a way to account for questions he now anticipates. He mentions, when describing the shape of the lens, "Some people think it's the distance and some people think it's the shape"—acknowledging that there are competing ideas and he cannot rule the other out. And he also talks about groups bringing attention to two "crossing points" in his group's diagram.

The process of multiple explanations and feedback from peers has challenged the group's ideas (is it shape or distance?), added new questions (what happens at the lens?) that other groups can help address, and suggested places where the group needs clarity. The students have an audience for their current and later work, and they have become an audience for others' work.

CHALLENGES

One of the most challenging aspects of entering a new discipline is the limited ability to notice nuances in different ideas. For example, students can look around the room at a variety of groups' whiteboards that all use the word *focus*, but have difficulty noticing that the diagrams are representing focus in different ways. Faculty need to model ways of carefully noticing, sometimes by simply stating, "It seems like we have competing views on our definitions of focus." Early in our semester, students often will tell us, as they look around the room, that the groups' ideas are similar. In these early moments, it is crucial for science faculty to point out variations in ideas. At the very least, modeling the practice of noticing reminds students to attend to the ideas with more than a passing glance. Over time, we find that as this practice of noticing is modeled, students demand more from one another: better diagrams, more precise language, and the use of agreed-upon terms. In fact, as the semester progresses, we can point to many moments when a student stops the class and says, "Wait, what do you mean by blurry?"

Another challenge, particularly early in the semester, is knowing what to say to a peer's ideas. Students need time to develop a shared language around feedback, modeled by the instructor and also by more capable peers. In Kayla's silent science example above, we saw explicitly how students try out the language that has been modeled for them: Leslie often writes on a student's draft, "So what I think you're saying here . . . ," as a way to preface her own attempt at understanding a student's idea. This way of phrasing a response honors the student's effort at meaning-making; furthermore, the phrase forces the writer and the reader to share the work of meaning-making. The phrase sets up the responder to make her own attempt at clarifying the idea as well. Simply saying, "I don't understand," puts all the work back on the writer. As with Kayla's responder, we often see language from instructor feedback taken up by students.

FEEDBACK AND GRADING

We do not formally grade the silent science or gallery walk activities. Instead, we use these structures as a way to reflect, revise, and deepen understanding of scientific ideas. Often, we end the activities by asking students to review their peer feedback and discuss the new ideas with their lab group. They reflect on their feedback and decide what changes they plan to make when constructing diagrams or writing an explanation in the future. Often the feedback spurs new questions. When students disagree, this is a rich opportunity to further the inquiry, through either new experiments or more discussion.

When we want a more extensive discussion about students' ideas and their feedback, we often will glance at the silent science papers while handing them back, searching for one or two that are particularly rich; if selecting just one, we look for a paper that has robust ideas and strong comments. Our goal is to have a discussion with students about what the writer was trying to show and how the feedback helps them further their ideas or improve the clarity of the ideas they are sharing. We also ask the reviewers of the paper to share what they noticed and we walk through the iteration of the ideas.

Another approach we use as an alternative to walking students through a piece of feedback is to highlight differing ideas. Students often expect that there is only one right way to explain a scientific phenomenon, and as they read through their peers' papers, they may wonder, "Should I have done that instead?" We find it useful to look at two papers that both offer good, but different, approaches. We look for two papers or whiteboards that have different ways of explaining a phenomenon, or where students have chosen to focus on different ideas. For example, in our class, one student might emphasize why the image is brighter when a pinhole is enlarged, while another might emphasize why the image is blurrier; their work will be quite different. By discussing their work and feedback (again by projecting images and having students walk through what they were noticing), we can launch a conversation about the variation in ideas and expectations for them.

Take-Home Messages

- In science, "the formal submission of a manuscript for peer review is not the first stage at which it has been subjected to peer scrutiny" (Harnad, 1990, p. 342). Similarly, in our classes, peer review happens throughout the course.
- Students need a chance to develop a language for feedback on scientific ideas. The instructor can model how to talk about ideas through written feedback and in conversations with students.
- Students need support as they learn to notice varying and competing ideas. Over time, pointing out competing ideas will help students see this as a valued practice.
- Students should be offered a variety of structures for giving and receiving peer feedback: orally through gallery walk discussions, on paper like silent science activities, online, and in table conversations in teams.
- As the ideas become more clear, the writing becomes more clear. Therefore, peer feedback in particular should focus more on scientific ideas than on sentence-level errors. Errors should be pointed out when they interfere with understanding ideas.
- Students need daily, iterative opportunities to give and receive feedback on scientific ideas, particularly before giving feedback on more high-stakes assignments. A major paper is hopefully not the first time they've commented on their peers' ideas.
- One goal of peer feedback is to give students an audience for their ideas. Our class functions in similar ways to a research lab; their peers (nascent scientists) are a real audience.

Reading Together

> What I think could be happening when [Newton] called those five colors "primary" was that when you shine white light through a prism, you see bigger/thicker streaks of the red, yellow, green, and blue. I don't think his use of the word *primary* is our same use of the word. —Amy

Other chapters in this book highlight students' writing within the classroom community: the data in their lab notebooks, the diagrams they construct, the whiteboards they share and critique, the peer reviews they perform as they read one another's ideas and vet them, and the more formal writing that happens as students summarize their findings. These inscriptions form the primary texts for our course. However, when we ask colleagues how it is they learn how to write something new—a letter for a tenure review, a grant proposal, a proceeding paper for a conference they have never before attended—they will, without fail, mention that they seek out high-quality examples from others. When we ask colleagues how they begin to work on topics that are new to them, they report doing an extensive literature review, summarizing and critiquing current research in that field.

So although this book emphasizes the role of student-created texts in instruction, which we use as our common texts in class, we know that it is important for students to develop other practices as well: how to reach beyond their peers to query the literature, how to critique and build on unfamiliar scientific work, how to place other scientists' work in conversation with their own ideas, and how to read an article to learn ways of structuring their writing, representations, and other stylistic choices. This chapter focuses on how we engage students in reading and using scientific texts as part of a broader scientific practice.

Of course, scientists don't approach the scientific literature as blank slates or without goals: We become interested in a field, topic, or technique somewhat tangential to our own, often beginning with a specific question as we move into a new arena. For example, in graduate school, Leslie, working in a physics lab on mesoscopic disordered materials (e.g., foams, crumpled foils, and sand piles), was introduced to a paleontologist through her adviser. The paleontologist's overarching question related to the microstructure of bone, which is porous and foam-like; he wondered whether the structure of bone could provide information on a dinosaur's size, weight, posture, and living environment, and had approached the physics department with these questions. Armed with a particular tool set from physics—ways of characterizing networks and forces in things like foams and sand piles—Leslie began to search the literature on bones to see how these tools could be brought to bear on an open question in paleontology.

In Irene's biology lab, there was a more gradual shift from research on the neural circuits underlying pain to those underlying motivation and reward. Initially, researchers in the lab were studying pain circuits and how opiates (endogenous and self-administered) affect those neural circuits. However, one graduate student started applying the electrophysiological techniques used for pain studies to examine reward pathways in the brain that were comparatively more complicated and less well understood. That work revealed interesting parallels between pain circuitry and reward circuitry in the brain. A grant was won and another grad student and postdoc joined the effort. This team of three introduced the rest of the lab to the literature by selecting an article each week for everyone to read and discuss at lab meetings. These lab meeting literature circles resulted in far-ranging conversations about open questions in the field of biology, experiments to try, and initial ideas about how what was known about pain circuitry could inform the growing data on reward pathways. Eventually the entire lab was reading literature on motivation and reward circuitry and pursuing research in that area.

In addition, while scientists may sit alone and read, this reading is nonetheless part of a much more social activity. In the first example above Leslie sketched out ideas in the margins of papers and in her lab notebook, shared these articles and ideas in physics lab group meetings, and discussed questions with the paleontologist and her adviser. She drafted a literature review, shared this with members of her thesis committee, and revisited the articles multiple times with a range of people. In Irene's biology lab, the graduate students' work and findings were shared in group meetings, papers were passed around, experiments were modified, and new grants were written to take into account the emerging research directions of the group based on this new literature. New papers were selected based on others' interests and were discussed by the entire group.

When scientists select and read a research article, the effort is generally goal-directed, related to ongoing work, connected to scientists' interests, and social. There are structures in place (some explicit, some woven into the fabric of the scientific community) to support reading, critiquing, and using scientific texts. Consistent with these characteristics, we introduce scientific articles only after students have gained familiarity with the phenomenon under investigation; developed their interests in relation to these ideas, questions, and nascent models; and have ideas and debates that the readings might address. Our job as instructors, prior to asking students to read about a topic, is to cultivate interests and questions in our students that the reading will speak to and build on. This is in contrast to typical experiences in science classrooms where students often are first introduced to a topic through a text and then asked to use the text to develop questions or replicate an experiment.

We also read "together" in class, sharing our questions, implications, and insights for our own work. We adopt structures that scientists use: sharing work online and engaging in journal club discussions of readings. A lesson plan detailing how we structure this is included in the online resources (at composingscience.com). Note that we never ask students to simply "discuss" a reading; we always have a defined "product" as an outcome of the discussion and a plan in place to facilitate that discussion. Without a product, conversations often are rambling or ephemeral. The product might be annotations, which are then taken up in lesson planning for the next day, or asking students to identify a quote that they would like to discuss further. In class, the discussions unfold around particular questions or ideas that students noted; we collect the answers or themes on whiteboards and in sketches, and we use them to articulate further questions. As in a journal club, students know why they are reading what they are reading, and have specific goals for the information gleaned from the article. The "facilitation" may feel contrived; after all, scientists don't implement structures like these when discussing a paper. Nonetheless, in a pedagogical setting with large numbers of students new to science, these structures actually can *facilitate* a more scientific, and less contrived, discussion. In their absence, we find conversations can be halting, students look frequently to the instructors for assessment, and a few voices dominate the conversation.

A brief aside: Scientists do not read articles only to inform their work. Much of the reading that scientists do involves reviewing and editing papers, and that lens—understanding how articles are reviewed and published—influences how we read everything: from a draft, to a manuscript, to an established part of the canon. The recommendations in other chapters, therefore, are also important for developing critical scientific readers.

How does a graduate student learn how to do this? To read papers for critical information, fill the gaps in his own knowledge, and find gaps that she might fill? To share these findings and solicit feedback from colleagues? In part, this happens in the context of the research group, where senior graduate students recommend articles to incoming graduate students. In addition, many science departments and medical schools organize journal clubs—settings in which students and scientists review an article related to their field, summarize its findings, and discuss implications. Reading scientific papers, a central part of academic work, is not a skill that graduate students are expected to possess; it is a skill that is taught throughout graduate school in implicit and explicit ways.

TAKEN UP

Below, we discuss two kinds of readings we use in our courses: (1) those that develop theory, often integrating models, experiments, data, and interpretation; and (2) those that provide data that might inform our own work. (We also include readings about the nature of science and applications of science to technology in our course.) In all cases, our set-up is the same: Students are introduced to a reading that responds to the ideas and questions that they have, and furthers those ideas. The readings are homework, and students read and comment on the text, sharing their comments with others in the class, usually via an online program such as Google Docs. These ideas are then taken up in class the next day, often first with small groups discussing particular sections. In this way, the authors and scientists from these readings become "members" of our classroom community and contribute to our ongoing discussions. These authors have a degree of authority, of course; scientists' ideas are not on an equal footing with student ideas. But students are nonetheless encouraged to wrestle with and challenge the ideas from these pieces. Below we describe one reading, from Isaac Newton, in detail, and then provide brief examples of other readings and how they are integrated into our course.

Developing Theory Using a Primary Text: Colors and Newton

We often start our unit on light and color by having students discuss the question, "Is every color in the rainbow?" This provides a segue into questions of light and color: How do the colors of the rainbow differ from other colors? What constitutes a "primary color"? Are the three primary colors primarily a biological or a physical phenomenon? Students generally come up with the idea that some colors are "blends" (e.g., brown) while other colors are "pure" (e.g., red)—and develop nascent theories about the role that pure colors play in generating blended colors.

In the class described below, students developed what they called the "cran-apple" theory of colors: Just because you can buy "cran-apple" juice doesn't mean that you can buy a "cran-apple" tree. Analogously, some colors we see (magenta, for example) can be produced only by combining other, more fundamental colors. Some groups begin to work with prisms, spectroscopes, and lights. Others work with paints or printed materials, where magenta and cyan are better than red and blue for making a range of colors. Surprised that red is not a good color for mixing to create other colors, one group describes red as too "strong." All of this complicates their initial ideas and early theories of primary colors.

At this point, after about a week of investigations, we introduce a reading from Newton, who first described the role of light in producing color.

In particular, we select "A New Theory About Light and Colors" (Newton, 1671/1672) and edit this to include the main points we want to discuss and to remove some of the length that is less relevant. (Editing is not necessary. In this case, we wanted to focus our conversation on particular claims. Faculty with other goals might include more text and ask students to determine the relevant parts.) We paste the reading into a Google Doc and duplicate that document so that each lab group has its own online copy. Students then are asked to read the document online for homework and add comments and questions and to respond to one another's comments and questions. We suggest that they translate difficult passages into "everyday" language and ask if others agree with the translation.

The document can be found in our online resources (composingscience.com) or at: tinyurl.com/TheoryOfColor. We encourage you to read Newton's words and work to interpret them before proceeding. You might even try adding a comment or responding to someone else's comment online.

In the piece, Newton opened by describing a puzzling observation: When a circle of light passes through a prism, the resulting pattern of light is not circular, but oblong, a hotdog shape. He noted that this shape was not predicted by the existing theories of refraction (which predicted that it would remain circular, just displaced from its original location). To explain the phenomenon, Newton began to think of rays as traveling in curved trajectories, and he ultimately determined, by passing a segment of the spectrum through a second prism and finding it does not further "split" into other colors, that white light is made up of different kinds of rays and these rays have a different degree of "bendability" (are "differently refrangible"). This effectively separates the circular spot of white light into many (overlapping) circular spots of colored light. Each color is bent at a slightly different angle, creating the oblong shape. Moreover, Newton argued, different colors are no more and no less than different "bendabilities" of rays.

In Figure 5.1 we see a screenshot from part of a Google Doc of student comments. One of the faculty has added comments after reading the student comments. There are five other documents, similarly marked up, one from each of the other lab groups.

Reading through these six documents, the instructors' goals are to figure out: (1) Are students understanding Newton's claims? (2) Are they seeing connections between Newton's ideas and their own? (3) Are they articulating implications for their own investigations?

In planning for the next day's class, we notice that some students are correctly interpreting Newton's claims, while others still have questions. Many are noticing that Newton's ideas relating color to the "bendability" of light have implications for their own theories of color. And we see many students drawing connections between the fact that red light is less refrangible and a class idea that red is too "strong" to be a good primary color. We decide to structure the next day's conversation to first discuss the experiment, then

Figure 5.1. Screenshot of Student Annotations of Newton's Text

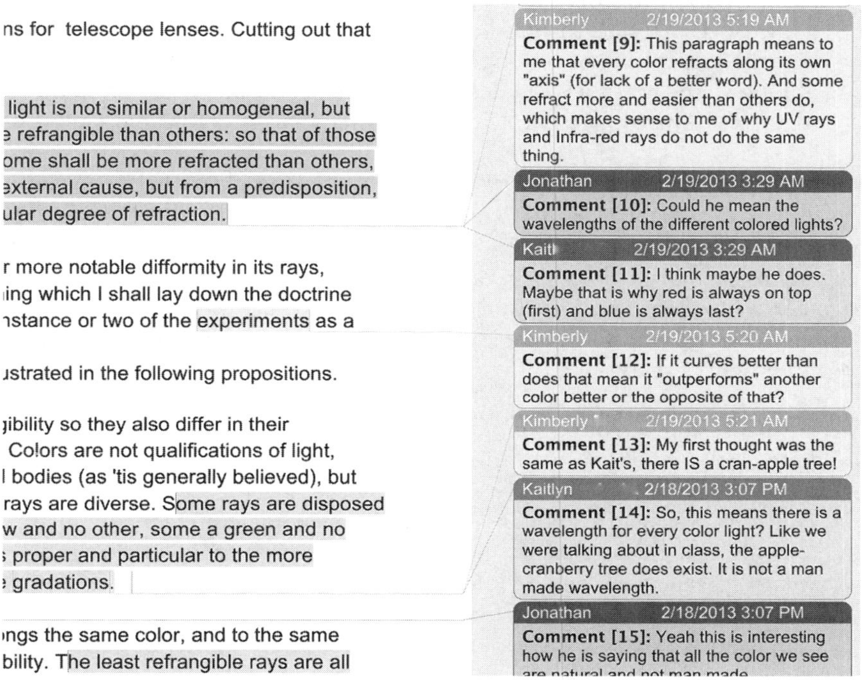

ns for telescope lenses. Cutting out that

light is not similar or homogeneal, but
ɘ refrangible than others: so that of those
ome shall be more refracted than others,
ɘxternal cause, but from a predisposition,
ular degree of refraction.

r more notable difformity in its rays,
ing which I shall lay down the doctrine
nstance or two of the experiments as a

ıstrated in the following propositions.

ɟibility so they also differ in their
Colors are not qualifications of light,
l bodies (as 'tis generally believed), but
rays are diverse. Some rays are disposed
w and no other, some a green and no
; proper and particular to the more
ɘ gradations.

ngs the same color, and to the same
bility. The least refrangible rays are all

Kimberly 2/19/2013 5:19 AM
Comment [9]: This paragraph means to me that every color refracts along its own "axis" (for lack of a better word). And some refract more and easier than others do, which makes sense to me of why UV rays and Infra-red rays do not do the same thing.

Jonathan 2/19/2013 3:29 AM
Comment [10]: Could he mean the wavelengths of the different colored lights?

Kait 2/19/2013 3:29 AM
Comment [11]: I think maybe he does. Maybe that is why red is always on top (first) and blue is always last?

Kimberly 2/19/2013 5:20 AM
Comment [12]: If it curves better than does that mean it "outperforms" another color better or the opposite of that?

Kimberly 2/19/2013 5:21 AM
Comment [13]: My first thought was the same as Kait's, there IS a cran-apple tree!

Kaitlyn 2/18/2013 3:07 PM
Comment [14]: So, this means there is a wavelength for every color light? Like we were talking about in class, the apple-cranberry tree does exist. It is not a man made wavelength.

Jonathan 2/18/2013 3:07 PM
Comment [15]: Yeah this is interesting how he is saying that all the color we see are natural and not man made

the implications that Newton drew, and finally to link those ideas to our own. (This is a common pattern when reading these kinds of texts.) Looking through students' questions and comments, we generate six prompts, one for each group. These are printed on slips of paper and handed to each group. Two prompts are related to understanding what Newton did and observed; two prompts are related to understanding Newton's interpretations of his results; and two prompts are related to ways in which students are noticing implications for their own work. We describe one prompt from each category below.

What Newton Did and Observed. Amy had a question related to the experiment: She asked about the representation and whether Newton saw or inferred the rays he drew. We decide that her group should recreate the "oblong" shape and look to see whether they can see rays of color traveling through the air. We generate the following prompt for Amy's group:

> In your group, Amy asked the following question: "I'm curious as to how he actually saw the colors. Did he see them in the air leaving the prism and going toward the wall (like we see a lot of representations)

> or did he just see the colors projected on the wall? And I'd really like to know what this oblong shape is. I can only imagine him seeing like a line of colors with the gradient of colors going down the line . . . is that the same thing as oblong?" Use a prism and tinfoil to create the oblong shape that Newton described (that is, just one prism) and try to answer Amy's question: Did Newton see rays of color in the air?

Newton's Interpretations of His Results. While the class had articulated the idea that there are "pure" colors and "mixed" colors, they were surprised by Newton's phrasing, describing the "pure" colors as primary: "The original or primary colors are red, yellow, green, blue, and a violet-purple, together with orange, indigo, and an indefinite variety of intermediate gradations." Kimberly asks, "Is he expanding what is considered the primary colors of light?" Amy (in a different group) responds to a similar question, saying, "I don't think his use of the word 'primary' is our same use of the word." We assign a group to look at the range of comments on this section and help clarify Newton's claim:

> Find this passage from the reading and review the comments on all 6 of the Google documents: "The original or primary colors are red, yellow, green, blue, and a violet-purple, together with orange, indigo, and an indefinite variety of intermediate gradations." Clarify Newton's claims and address the questions that were raised online. In particular, is there evidence in the text that Newton thought of primary colors the way that we do?

Implications for Students' Own Work. As noted above, we had a lively discussion in our class about colors in the rainbow. In particular, some rainbow colors appear to be "blends"—orange is a mixture of red and yellow. Does their appearance in the rainbow mean they are pure colors? Or are they blends? An analogy to cran-apple juice was drawn (is the color orange like the flavor cran-apple—a mixture of more fundamental flavors?), and students refer to blended colors as cran-apple. Many students believe Newton made a claim that supports the idea of a "cran-apple" tree. "The apple-cranberry tree does exist!" "So, this means there is a wavelength for every color light? Like we were talking about in class, the apple-cranberry tree does exist. It is not a man-made wavelength." "My first thought was the same as Kait's, there *is* a cran-apple tree!" We assign a fifth group of students to address whether they find support in Newton's text for the idea that a "cran-apple" tree exists:

> Many students believe that this reading endorses the idea that a "cran-apple" tree exists. Find support in Newton's writing that this is true—or not true.

Each group is given a prompt at the start of class. Students have access to laptops to view the annotated readings that each lab group has created, so they can read the questions and comments other students have generated as they prepare a whiteboard to share with the class. The instructor circulates among groups, addressing questions and helping groups to focus their main ideas for the class conversation. Because the class has access to everyone's comments and questions, students can refer to one another as they discuss the paper: "Trevor asked . . . and Andrew thought perhaps . . . but we disagree, and think that Newton meant . . ."; or, "We liked how Kait phrased this . . . and we want to add . . . ." This invites other students into the conversation, and highlights the social nature of reading in science.

As a whole class, we progress from articulating the rationale for the experiment, asking the first group to explain its observations of the oblong shape. The instructor clarifies that the laws of refraction suggest that the circular spot of light should stay circular, just bent off to one side—the fact that it "stretches" told Newton that something else was going on. A second group discussing the findings then describes what happens when a second prism is used to further split the rays. These findings are related to the first group's discussion. We then move through the remaining prompts.

The role of the instructor during this conversation is to facilitate a discussion and prevent it from being a simple "reporting" of what each group determined. Some semesters, this is not a challenge as students quickly engage in a conversation. Other semesters, we might need to prompt them: "I know that a lot of people had a question about this part—does that make sense?" Or, saying to a group that isn't presenting, "I know your group was struggling with that question, too—this group is saying that 'cran-apple' doesn't really exist, right? So do you agree?" Or, "You know, I think this is related to what Amy's group just saw with the prism—can you share a picture from that experiment?"

This preparation and conversation generally takes the entire class. If there is time at the end, you can give students time to collect their thoughts in their lab notebooks or, as a lab group, discuss their next steps.

This discussion mirrors the work that a journal club might do, but deliberately structures the class so that students are presenting on areas where they have the most confidence, and organizes the conversation to progress from data to interpretation to implications. Because we have each group respond to a question that another student in the class articulated, the reading is strongly tied to the questions, ideas, and implications of other students.

The following are some benefits to the selected reading that we use as an example here: First, it is primary literature—this is the first account of the relationship between white light and color that is consistent with our current understanding; that is, that white light is composed of colored light. Second, it is a "complete" scientific story: Newton recognized a puzzling observation—puzzling because it ran counter to existing theories—theorized a mechanism to explain that observation, and then invented a novel experiment to further explore that mechanism. He then spelled out the conclusions he drew from his experiment. (He also recognized some fascinating implications from this new theory—namely, that chromatic aberration is an inherent problem of lenses. We cut this out, but it would be interesting to include in a class that was examining lenses.) Third, it is relatively brief; we select just over 2 pages to read together and the entire reading is not more than 10 pages. This allows for a focused conversation. Fourth, it is not entirely correct: The distinction between "original" and "compound" colors is not as tidy as Newton supposed (certain shades of yellow, for example, can be "original" or "compound"). Letting students know that not all of the ideas in the paper have held up over time helps them to read it skeptically. Finally, the reading speaks to the work that students are doing (in particular, the idea of "homogeneal" and "heterogeneal" colors is useful), but does not answer the questions (some of which can be seen in Figure 5.1) they have been directly engaged with, such as questions having to do with colorblindness, primary colors, why red occupies so much of the spectrum, and the relationship between computer pixels and newsprint stippling.

From their comments and the ensuing conversation, we can see that students are not treating this text as authoritative statements to memorize, but are interacting with the text in a scientific way. They wrestle with and challenge some of the ideas that Newton presented; they suggest that what he meant, for example, by "primary" might not be the same as what our class means by primary; they consider whether his descriptions of red have implications for their questions about primary colors. In a sense, it is a way to bring Newton himself into the classroom conversation, not as an authoritative figure with the "right answer," but as a fellow scientist with whom one can engage in conversation.

As a final note, scanning across the annotated documents, selecting prompts, and assigning those to groups involves no more than an hour of preparation on the part of the instructor. This includes grading student comments (credit/no credit) and generates the lesson plan for the subsequent day.

Interpreting and Using Data from the Literature: Case Studies

Scientific articles on experimental work will, with rare exceptions, include a thorough discussion of experimental procedures and results. This often is described—particularly when introduced in K–12 settings or as a science fair requirement—as facilitating replication.

However, few experiments are actually replicated. There are many reasons why: The experiments may represent years of work, modern science is expensive, and even when time and money are not prohibitive, funding is rarely available to re-examine accepted experimental conclusions. Or, despite rigorous and well-documented methods, the experiments may not be replicable: They might take advantage of a particular phenomenon such as an earthquake, outbreak, or supernova. And, perhaps most important, replication was never the point of rigorously describing methods. Instead, as Feuer, Towne, and Shavelson (2002) note, it is part of the scientific community's

> concerted efforts to train scientists in certain habits of mind: dedication to the primacy of evidence; to elucidation and reduction of biases that might affect the research process; and to disciplined, creative, and open-minded thinking. These habits, together with the watchfulness of the community as a whole, result in a cadre of investigators who can engage differing perspectives and explanations in their work and consider alternative paradigms. Perhaps above all, communally enforced norms ensure as much as is humanly possible that individual scientists are willing to open their work to criticism, assessment, and potential revision. (pp. 29–30)

To focus students' attention on the work of evaluating, critiquing, interpreting, and using experimental information and data, we will, at times, extract part of a paper that covers the experimental methods that were used. Students do not repeat the experiments or observations, but instead make sense of how the particular experimental methods led to a set of data and conclusions, or they may use the findings to inform their own work, or employ the methods to develop their own investigations. We also remove the scientists' interpretations so that students can interpret the data on their own.

One example, again from our unit on color, is a set of descriptions of people who are color-blind. These do not explain the mechanism behind color-blindness, but simply describe, in vivid detail, the types of color-blindness that have been identified, the colors that someone would see, and which colors are confused, along with information on the physiology behind color-blindness when possible. Unlike the other readings, these are edited from published case studies in the medical literature or popular scientific texts (such as Oliver Sacks's books) to present summaries of the data.

CHALLENGES

When we think of how scientists select articles to read, it becomes clear that this is a skill that is acquired over time as a member of a scientific community. We learn from our colleagues, from citations, and from our own experiences with publishing, which journals provide strong review articles, which ones offer resource papers, where to find articles on instrumentation and other experimental techniques, and where to turn for "last word" papers that summarize a field of inquiry. We know which conferences have rigorous review and associated proceedings. We have a sense of which research groups we like to read—whose work is close enough to our own that we try to stay abreast of their publications. And we have colleagues who direct us toward papers we otherwise might not have found. All of this is to say that while many undergraduate writing courses teach students how to write "research papers" by using search engines and related tools to find primary literature and write a literature review, this is a very low priority in our course. In our experience, the work that students do when writing these literature reviews and research papers mimics scientific research in only the most superficial of ways. Instead, we prioritize texts that students can use in scientific ways: to inform their own research questions, to critique in light of their evolving ideas, and to discuss with their peers. In many cases, particularly in our course for nonmajors, these are not journal articles and students are not asked to find them; instead, the faculty selects texts to read.

The challenges we face when selecting, reading, and discussing texts are: (1) finding appropriate readings; (2) encouraging students to read them critically; and (3) structuring class discussions of the readings so that they are generative and not limited to, "I agree/disagree" or, "I liked this part." In large part, concerns with items (2) and (3) are mitigated when we find a particularly good text. The techniques described in this chapter (using Google Docs or similar technology, structuring the class discussions to attend to students' questions and ideas, and focusing discussions to provide a product) are ones we've used to promote critical reading and generative conversations.

In short, our advice is to select texts that students can make use of in scientific ways: texts that they can evaluate and critique even as relative novices; that they can use to inform their own work; that speak to ongoing questions or present a novel representation or data; and that present themselves as models of writing that demonstrate a "dedication to the primacy of evidence; to elucidation and reduction of biases that might affect the research process; and to disciplined, creative, and open-minded thinking" (Feuer et al., 2002, p. 30).

These texts have come from a range of sources. In Irene's biology courses, current articles from research journals on animal behavior were accessible to undergraduates. Technical articles on color were less accessible, but not impossible. In particular, students became interested in the then-new, four-primary "Quattron" technology, and making sense of the diagrams in these articles was challenging but productive. When examining light and the eye, we drew heavily from Oliver Sacks and his detailed descriptions of visual disorders. When working with a group of science teachers, we offered them conflicting definitions to reconcile that were drawn from science textbooks, Wikipedia, the Merriam-Webster dictionary, and a scientific review article. Popular articles from *Scientific American* and *Science News* have been incorporated. And, as described above, early work by Newton and Galileo has been useful for physics topics.

We rarely read more than a few pages at a time, particularly for very technical pieces. Even when we read longer texts, we assign smaller sections to help focus students' discussions. Again, while this may not have a direct analog to scientific practice—scientists in a journal club or other seminar typically will read and discuss an entire paper—we find that the conversations have a focus to them when we give shorter readings, and that these conversations are more productive.

Occasionally, however, students need support in understanding how to comment—they need guidance on how to read and query texts for this class. If you feel that this is the case (and not that the text itself is the issue), first examine the overall structure of the course: Do students have opportunities to interpret, assess, and critique ideas throughout the course? If not, they might be reading the text the way they read a textbook: not critically, but to find key phrases they should "know." In this case, you might help reframe their work. Start with groups working at tables, reading and commenting online (using Google Docs) on a shared computer. In addition, print a copy of the text for every student; be explicit in describing why you have chosen the text. Explain how it relates to ongoing questions and how it adds another voice to the conversation. Have a laptop for every lab group and project the Google Doc for all to see. Ask students to highlight any sentences they have trouble understanding and instruct the teams to try to interpret the highlighted sentences in the comments. We find that, as students see the document being populated with questions, interpretations, and discussions, little support is needed beyond that.

Finally, from "think, pair, share" to "fishbowl" to "Socratic seminars," there is no shortage of advice and tips for structuring and facilitating classroom discussions. We take advantage of a range of these structures; when done in concert with a well-chosen article and with preparation (via preclass assignments, like those using Google Docs), these can help create very productive, lively conversations.

FEEDBACK AND GRADING

The students provide much of the feedback: They comment on one another's ideas in the text, they agree or disagree with comments, and they give responses to questions that others raise. The follow-up discussion in class the next day offers more opportunities for feedback on the ideas and questions that the text generates. This feedback does not stand alone: This is a course where students frequently comment on one another's ideas in lab groups, in whole-class discussions, via whiteboards, and on homework. The scientific texts provide one more perspective that students respond to, another set of ideas that they participate in assessing.

As instructors, we weigh in on the conversations as well, both online and in class. For example, in the example from Newton, we answered questions about definitions (e.g., the meaning of refrangible). And by selecting some of the questions and conversations for further discussion in class, we highlight particular ideas as worth unpacking further.

To grade their work, we typically give a credit/no-credit grade for the reading assignment. We ultimately hold students accountable for the ideas from the reading: If they later suggest ideas that contradict the reading, they need to be able to explain why they disagree. The ideas from the reading become part of the set of ideas that the class uses.

Take-Home Messages

- When choosing articles or other texts to read, we keep the following in mind:
  - The priority is on selecting texts that students can interact with in scientific ways. This, we believe, is more important than selecting texts that practicing scientists read.
  - As in other professional settings, when scientists read others' work, it is generally a goal-directed and purposeful activity, related to their own work and interests. Similarly, we ask students to read texts related to their current work and questions. This requires that readings be introduced after they have engaged with a topic and formulated tentative ideas and questions.
  - You might select a reading for students to use to analyze the structure of an argument; in this case, a complete article may make the most sense. On the other hand, if you want students to consider a particular representation, claim, or experiment, an excerpt may help generate a more targeted, focused conversation.
- Scientists read together. As reviewers, editors, in journal clubs, and in lab groups, scientists discuss and make sense of scientific writings with one another; similarly, readings for class should be part of the social meaning-making that students do as they consider scientists' ideas and implications of those ideas.
- Online technologies can make "reading together" relatively easy: Google Docs and Medium both offer straightforward methods for commenting on text.
- When discussing a reading in class, having a "product" for the discussion can help provide structure and permanence to the conversation, and help novices have a way to make sense of the text and discussion. By "product" we mean things like developing a list of questions that the text raises, answering a set of questions that came up while students were reading, comparing scientists' ideas with the class's ideas, repeating an experiment that is hard to understand, or describing implications of experimental data for their own models.
- When discussing a reading, particularly in a larger class, facilitation is important. Give some forethought to how you will have the class make progress on the product(s) for the discussion: small groups, partners, and so on.

Homework

> I was able to just write my thoughts and ideas without being afraid if they are right or wrong. Almost every other class I have had, I never had any wiggle room to express why I was thinking a certain way. It was either you're right or you're wrong, no in-between. I really like how I had that freedom in this class. —Pamela

Looking back over the previous chapters, the parallels between students' work and scientists' work are obvious: Scientists use notebooks, share ideas informally on whiteboards, construct diagrams and revise and refine them over time, engage continually in giving and receiving feedback on ideas, and read, using the literature to inform their work. We advocate for incorporating those literacy practices in science instruction, so that classrooms are not places where we "play school" but sites where students develop their scientific ideas in scientific ways.

Our students, however, unlike scientists, are assigned weekly, graded homework. And while scientists have a range of short- and long-term deadlines, there really is no direct parallel to this very "schooled" practice of assigning the same homework to the entire class every week. Nonetheless, homework is a critical and meaningful element of our course. The connections to scientific practice are not as direct, but homework can be considered a parallel to the ways in which scientists assume responsibility for collegial contributions to their professional communities. When we, as scientists, gather at a lab meeting, committee meeting, conference, journal club, or writing group, or participate in medical rounds, there is a degree of preparation that happens prior to these gatherings. So much of scientific work is a social process, but time together—as scientists and as students—is limited; it is a part of our shared responsibility to the group that we show up prepared.

This requires, of course, that the homework is, in fact, a preparation for our course and necessary for participating in our class. Traditionally, homework in science courses is back-of-the-chapter problem sets, which serve primarily as demonstrations to the instructor that the student has learned the requisite material and is ready to go on to the next step. When a student does not do her homework, or does it poorly, in traditional classroom settings there is very little effect on the class as a whole. The lecture proceeds as planned or the lab is completed without the student's input. The student who fails to do the homework likely will have difficulty keeping up, but there is no shared responsibility to the class that is unmet. Our homework, as will be described in more detail below, is part of the emerging narrative of our course: It picks up on ideas, debates, and questions that have been introduced in class, and furthers them, asking students to weigh in with their ideas, anticipate and respond to counterarguments, develop novel representations, or explain a puzzling phenomenon.

Rarely, then, do we simply collect and grade the homework. Often we assign homework over the weekend and begin class on Monday by having groups discuss the assignment. Their responses are the raw material of our curriculum as their ideas are shared on whiteboards, in silent science, or as a master class, or picked up in small groups as they investigate claims that have been made. We highlight different students' ideas, using this dissent to motivate further investigations. Alternatively, we may collect the assignments first and read through them, seeking themes in the ideas that we can use to make instructional decisions. That is, we do not read evaluatively ("are students getting this?") but instead read to understand the landscape of ideas that are in the classroom, seeking opportunities to engage deeply with every student's ideas. Both of these techniques are described in the following section.

TAKEN UP

With some exceptions, we write the homework assignment after the end of class on Friday and post it online that evening; it is due at the start of class the next week. Below we share two assignments that offer ways in which we use student ideas to generate homework, and, briefly, how we then use the homework to further our instruction. The first, on color theories, presents a range of different theories that students have developed, together with their observations, and asks students to consider which theories are consistent with which observations. The second assignment, on the lens, presents the findings and models resulting from the work of lab groups who have been tackling different questions, and asks students to use those findings and models to explain a phenomenon.

Color Theories Homework

In this class, we began the semester by asking the question, "Is every color in the rainbow?" After our initial conversations, which raised questions around whether some colors are "pure" but others "mixtures," and the role this might play in determining which colors are "rainbow colors," we gave students simple spectroscopes: a tube with a diffraction grating in one end that essentially makes a rainbow out of light. When students looked at sunlight, the tubes generated rainbows that looked "complete." But when students looked at other colors, or at white lights covered with colored filters, they noticed fewer colors present in the spectra. (As faculty, we deliberately selected six specific filters because of their significance as primary colors, but students did not yet know this.)

Nascent theories were developed to account for what they saw, and these were discussed and named in class. For example, one student, Kristin, had the idea that if two colors blend to produce a third color, the way that yellow and blue colors mix to make green, then there must be some green present in those "parent" colors. She expected that when one looked at yellow and blue light in a spectroscope, the resulting spectra should show some green. The class named this the "offspring theory." Other theories they developed and named were "the hot sauce theory," "the midpoint theory," and the "single-color theory."

For homework, we asked the students to consider each of the four theories that they had developed (Figure 6.1). We generated an online survey that asked students to determine which observations, pictured on the homework, could be explained by which theories, and which observations, if any, needed a new theory to explain them. After each picture, they not only selected which theories could be used to explain the data, but they also had to enter an explanation for their choice(s). This, we expected, would force students to carefully consider the theories they had developed—where they were too unclear to support careful predictions, we would work toward clarity; where students had varying interpretations, we would work toward precision; and where the data could not be explained by an existing theory, we would work toward new ideas.

For example, some of the explanations of the color of a red + yellow filter (which appears only red, and not the orange that many students predict) were:

1. The hot sauce theory applies to the combination of red and yellow filters. When combining red (pure capsaicin) and yellow (habanero chili), the red (pure capsaicin) will always be more intense than the yellow (habanero chili). . . . The "force" of the greater will always dominate over the lesser.
2. The hot sauce theory and/or the midpoint theory could be used for this image. Either the red filter is stronger than the yellow filter, and therefore shows up instead of the yellow. Or instead, we could say that the slightly darker red produced by the yellow and red overlap could be their midpoint color.
3. I feel like we need a new theory on this one because our current theories don't work for this. Some may argue it's the hot sauce theory but I don't think it works for this one. I feel like the red

Figure 6.1. Homework: A Survey on Colors

Several ideas were introduced on Wednesday to explain what happened when filters were placed on the overhead and the color examined using a spectroscope ("rainbow maker").

- The hot sauce theory (mentioned by Emma):

  Colors closer to the "red" end of the spectrum are "stronger" colors. If we look at a gel through our spectroscopes and see that it includes, for example, green and red, it will appear red because that end of the spectrum is stronger.

- The single-color theory (mentioned by Kristen):

  The reason we see a range of colors in our spectroscopes is because of ambient light in the room. If viewed in ideal conditions, a red filter will only show red light.

- The midpoint theory (mentioned by Kristen, goes with the preceding point):

  Each color is its own wavelength (the number on the square-shaped rainbow maker). If you mix two colors, you will see the midpoint.

- The offspring theory (mentioned by Kristin):

  If a color is "made" from another color (the way that yellow + blue makes a green baby), then that first color must "have" the other color in it (so yellow and blue both should show green when using the rainbow maker).

Over the next few weeks, you'll develop additional theories, or improve upon these theories, as you collect more evidence and data. The question in this assignment is whether these theories can account for what we have already seen. You'll be asked to decide whether, for example, theory one (the hot sauce theory) is consistent with the observation that, say, sea blue filters show the GBIV part of the spectrum. Check boxes on the survey (link is online) and explain your thinking.

filter is such a dark shade of red that it won't let the yellow come through. I feel like if they were both even shades that the yellow would have a chance to shine through and show orange.

4. This again may have more to do with how the brain interprets color; there is perhaps some benefit for picking out red.
5. "Offspring" makes sense but when we combined those filters, an orange baby was not made so we need a new theory.

Online surveys are straightforward to set up, and it's easy to analyze student responses just before class. Where students differed on answers, we knew we needed clarity. There were theories that explained very few observations, and those we decided to abandon. And some theories (midpoint, for example) worked well for some colors and poorly for others, suggesting it was promising but incomplete. Some students, we noticed, disagreed with a theory regardless of its predictions. (This is not entirely unreasonable—it's hard to imagine a physical model for color that would make sense of the hot sauce theory.)

In this way, students see their ideas being taken up in the assignment, and completing the assignment is necessary preparation for the next class meeting. From students' explanations of their choices, the instructor can make decisions about next steps to take and about how to leverage ideas that students have introduced. That is, the homework is an integral part of the emerging ideas in the class, and participating in the homework facilitates the ongoing development of ideas.

Lens Homework

By the time we were nearing the end of our unit on the eye, lab groups had developed a range of observations and models relating to the way that light interacts with a cornea (the curved outer surface of the eye) and lens. While these ideas had been shared with the whole class on whiteboards, these ideas had not yet been incorporated into a shared model for how the eye interacts with light to create a focused image. In our homework, we took images from the whiteboards and synopses of different groups' ideas, asking students to use these to explain the implications for their findings on how an eye creates an image of an object.

The homework began by summarizing the findings and models from three groups, together with images from their work:

- If we shine a Maglite through a lens, we might see a blurry circle of light or—if we get things just right—a bright dot of light (Figure 6.2A).
- If we shine two Maglites, and they're the same distance from the screen, you can get two dots side by side, but "reversed" (Figure 6.2B).
- Light travels in straight lines most of the time, but it bends when it enters gelatin (Figure 6.2C).

We then told students that these observations should help them think through the week's assignment, and that their job was to consider what happens when a nine-bulb flashlight shines into an eye that has an iris (with pupil) and a lens behind that pupil.

When students met in class on Monday, we began by having groups develop a whiteboard to summarize their ideas. These are described in more detail in Chapter 4. That is, their responses were used immediately in the next class, as they shared, refined, and developed more precise ideas about how the eye constructs an image on the retina.

Figure 6.2. (A) A Diagram Showing That Rays Through a Lens Cross; (B) A Diagram Showing That Images Are Reversed; (C) Using Lasers Through Gelatin to Show That Light Bends When Entering the Substance

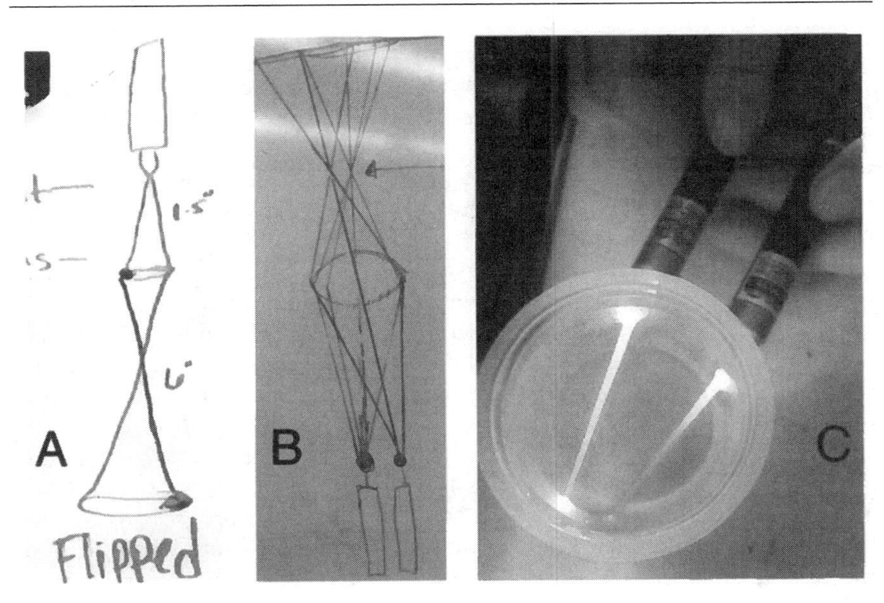

These two examples emphasize how the homework assignments, drawn from student ideas, are necessary for participating in our class. Ideas from students are woven into the curriculum, as these ideas are used to generate assignments and then taken up as we discuss and refine their ideas from homework. Later in the chapter, we provide an instance of how we use homework to better understand and draw out ideas that students are developing.

CHALLENGES

Students are familiar with problem sets and worksheets that are summative, and our open-ended homework asking about preliminary ideas can be disorienting at first. The homework submitted by students varies widely in length and care; some students are concerned that they should "know this," and others sketch out ideas cavalierly. As we use ideas in class, the goals of the assignments become more clear, and students recognize that the homework is an opportunity to reason carefully through their own ideas before working through the ideas with their peers. By seeing the work that their classmates do as we discuss ideas, students quickly develop expectations around homework. We, of course, still see students working on their assignment just before class, and there are always students who do not complete the assignment. Our approach here is never to shame a student: We acknowledge it matter-of-factly.

Another challenge lies in drafting the weekly assignments. As we teach, we notice, collect, and curate student ideas, smartphone in hand. After class, the instructor jots down notes from the day, taking 10–15 minutes to reflect on students' ideas, struggles, and any emerging themes. For the first homework described above (shown in Figure 6.1), we recognized that the hot sauce theory appealed to students in that it "explained" a lot but was too vague to use predictively and not connected to any model of how a particular color of light might be stronger than another. We knew there were enough data in the class to winnow out some ideas, and we used the homework to force conversations around this topic: that our theories were not clear enough to be predictive, and that data could constrain the theories we were considering. In other semesters, we might notice one particularly compelling idea and ask students to engage with that idea, as applied, say, to their own data. The real challenge for the instructor is to carefully attend to students' ideas while also having an eye on the "horizon" (Ball, 1993)—considering ways in which students' ideas can progress to become more scientific, and what kinds of questions and conversations will lead us there. This orchestration is both the greatest challenge and the greatest joy in teaching this class.

Because homework is a part of developing ideas in class, students should receive feedback quickly. When possible, the ideas are used immediately in the next class, but at times we first collect them (as described below) and quickly read across them to identify themes and develop instructional responses to them. For this reason, sometimes we use online methods (such as the survey). Other times we read through their ideas and craft our instructional response quickly, but wait until later to finish the grading and individual feedback.

FEEDBACK AND GRADING

Feedback

There are two ways we think of feedback on homework. The first is how the ideas are responded to publicly, in class. This kind of feedback that students receive on their ideas is threaded throughout this book: They use their ideas, often generated in homework, to create silent science, gallery walks, and whiteboards. While strongly shaped by the instructor, this feedback comes primarily from peers. The second is the responses that students receive from the instructor on their assignments. Here we are not only crafting the class narrative, but also recognizing and supporting the varying individual narratives. For example, we reach a group consensus on how the eye works, but students have become interested in very particular questions along the way.

This individual feedback takes the form of a brief letter to students, where we call attention to things they are doing well, questions we have, and what we are noticing in their work. This is an opportunity to notice the themes that are emerging from the work of individual students as they engage in "a reflexive process of transforming scientific discourse in a way that is authentic and personal" (Levrini, Fantini, Tasquier, Pecori, & Levin, 2014, p. 93). For example, Kaitlyn has been diagramming what happens when an eye has no lens, and she attends to a question that few are noticing: The back of the eye is curved, and, like a curved mirror or screen, she wonders whether this will distort the image. She also has a unique way of shading her diagram to show intensity. Feedback from the grader, Leslie, calls attention to things Kaitlyn is doing well, recognizes the unique perspective she takes, and points her to other resources:

> Kaitlyn,
>
> You're doing lots of things in your diagrams, and you're choosing how and what to diagram based on the effect that you want to highlight. It looks so great.
>
> I'm seeing three different things you're showing:
>
> 1. The image in the retina is smaller than the object. (This is good b/c, if I want to see, say, a tree then it must "shrink" to fit on my retina.)
> 2. Objects on the "edge" of our field of view are smaller than objects in the middle of our field of view, like a fun-house mirror.
> 3. Objects are blurry—in fact, from your diagram it looks like one "Seurat spot" will send rays to a region on the retina that's bigger than the hole in the iris.—I *really* love how you're showing this with the shading. It actually looks a lot like this (and seems to be a similar phenomenon): www.stargazerpaul.com/solar%20eclipse%20diagram-pderrick.jpg
>
> I'm really intrigued by the question of "distortion"—the fact that the edge is shrunken while the center is not.—It reminds me of this (which I saw just yesterday), where a teacher has students looking at why distant objects look smaller: noschese180.wordpress.com/2013/05/07/day-146-size-distance-and-eclipses/
>
> Overall, this is great (of course)—so, so excited by the work your group is doing!

In this class, most students had been describing points in the world becoming larger circles of light on the retina and noting that this causes blurriness. Esmerelda had a different (but not inconsistent) way of describing blurriness, where each point on the *iris* projected a single image on the retina, and these images overlapped to cause blurriness. This idea had come up earlier for her on an exam. Here, the instructor notices this, refers to the exam, and recognizes this theme in Esmerelda's work. In addition, she explicitly identifies this as a problem the lens will solve.

> Esmerelda–
>
> I love how this builds from your midterm—thinking about how a large pinhole causes there to be multiple images, all overlapping. Most people are describing this as the "Seurat spots" each becoming too big, while you're saying (I think) that each Seurat spot makes more than one image. Pen diagram is awesome.
>
> This seems like a challenge that the parts of the eye must solve—how to take those multiple images and turn them back into one.

In the final example of individual feedback, Trevor has created a numbered list of ideas, leading the reader step by step through the main points of his model. The instructor identifies approaches that he takes that are helpful and then offers a suggestion, including annotating his diagram (see Figure 6.3) by snapping a photo and using simple software to add suggestions:

Figure 6.3. Instructor Annotations on Student's Diagram

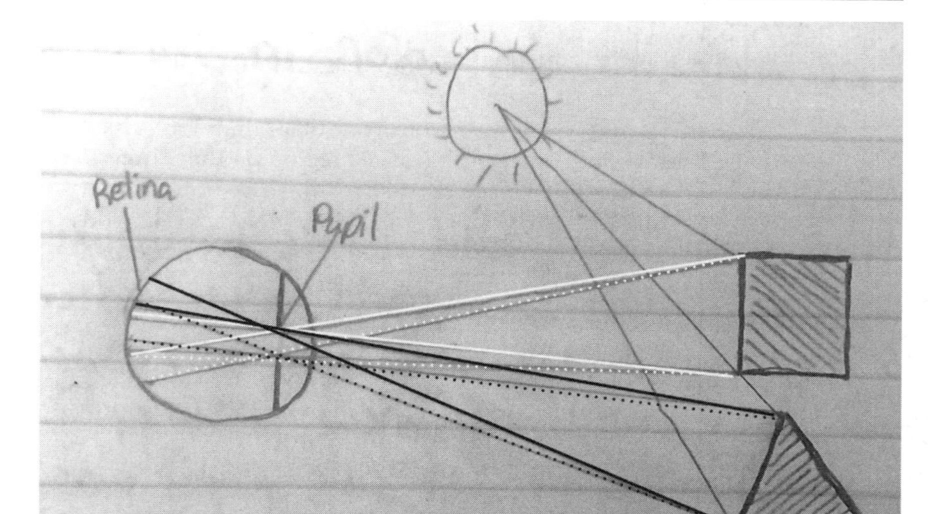

> Trevor,
>
> I like the numbered list—reminds me of Berkeley's theory of vision (where he explains depth perception through a series of numbered ideas).
>
> I really like this phrasing: . . . "images would get from a single point on the objects to multiple points on the retina." I think you could do more in the diagram to illustrate that—right now it looks like a single point expands to be a large area and not a multiplicity of overlapping images. Maybe coloring or dotted lines could help show that more clearly?

This instructor feedback also provides a model for the feedback students offer on one another's writing, that is, recognize what the author is trying to do and help support the author in doing that work. This is not an opportunity to simply correct an idea or make a diagram that is more in line with what the instructor expects.

A final form of feedback that we offer on writing is to look across all the students' homework and identify the range of themes and ideas that have been presented. Much of the time this is something that an instructor does on her own and is part of the planning for class—we identify themes and conflicting ideas that we can subject to scrutiny. At other times, however, we summarize the themes for students or introduce the varying ideas that we see and ask the students to help summarize them.

For one lesson, Leslie summarized the themes from the homework:

1. There is too much light.
2. There is too much information or signals.
3. There is a lack of "organization" to the light falling on the retina.
4. Every point on the retina gets light from every point in the world. Every point in the world sends a ray of light to every point on the retina.
5. There is an overlapping or adding of light rays when they land on the retina.

She expected these ideas to emerge in whole-class conversation and ultimately imagined a discussion that would articulate how these different ideas suggested different kinds of problems the eye must solve. These are Leslie's planning notes:

1. I need something to "dim" the light; I might be able to see if I was in a very dim room. It's like leaving the shutter on your camera open for too long or staring at the sun. Or it's like someone screaming something *so* loudly you can't hear what they're saying.

2. It's like being at a party and trying to hear a conversation—it's not that the person you want to hear is mumbling or quiet, or that they're too loud, but that all the other voices drown them out. This suggests that there is an image on the white paper, but it's somehow overwhelmed by all the other light.
3. The light needs to be somehow "organized"—not necessarily blocked or dimmed, but rearranged or regrouped. It's not like being at a party with many different voices and it's not like one voice that is just too loud, it's like a million people all saying the same thing but not in sync.
4. It's a "line of sight" problem. Without a pinhole, all objects are in the "line of sight" for all points on the retina. (I can't figure out a good "sound" analogy for this one!)
5. This ties together our ideas about light with why we wouldn't see color—b/c we see all colors at once. (So it would be like hearing multiple sounds at once and hearing the "chord" instead of just a pitch.)

In this way, the range of ideas taken from the homework is available to all students. The work the instructor has done of curating the homework offers students an opportunity to notice and analyze the varying scientific explanations.

Grading

Grading homework is trickier: students are *developing* their ideas, so we primarily grade whether students have carefully and thoroughly considered and responded to the homework. But there are also ideas from class for which students should be accountable, and our grading will reflect the correct use of these ideas: definitions we have constructed, models that we agree on, or data that we have established. By using a limited grading scale (three points for simple homework, like completing a survey, and five for more complex assignments), we find that we can grade relatively quickly and in a way that honors developing ideas and established ideas. A perfect grade, indicating that the student has been thoughtful, thorough, and responsible to the claims we have made in class, is somewhat rare. Much more common is the "B" grade, where the student could have taken more time or considered class ideas more thoroughly. If a student wants to discuss their grading, we will have them look across a range of assignments—some receiving perfect grades, some lower grades—and discuss the differences we notice.

Take-Home Messages

- Homework, like all writing in this course, does not stand apart from the ongoing work students are doing in constructing ideas. It draws on the ideas that have been developed and is used to further refine those ideas.
- Homework is a way for all students to participate in the community of scientists: It serves as a way for students to engage deeply with other students' models and data, and is preparation that students do for participation in class.
- Because homework is used in class, feedback—whether from peers or faculty—must happen quickly. This can be accomplished by immediately using the homework in structures such as silent science or whiteboards; or faculty can quickly review the ideas as a formative assessment, using them to inform instruction.
- Instructor feedback models for students how to offer their own peer reviews: attending to the author's ideas and not one's own.
- Grading and feedback are not the same: When possible, we grade quickly, but this is far less critical than rapid feedback.
- Because our homework is rarely summative, but instead is embedded in an ongoing and iterative process, it positions writing as a way to learn and not simply an assessment of learning. Feedback and grading reflect this.

Part II

WRITING TO COMMUNICATE

Formal Writing Strategies That Share, Critique, and Defend Scientific Ideas

CHAPTER 7

Definitions

> But we just made it up. It's not like scientists just made it up. Wait. Scientists just made it up! —Gregoria

The previous chapters described efforts to *develop* ideas through reading and writing. Assignments and inscriptions were understood to be drafts—not only in terms of writing, but also in terms of the ideas they conveyed. In Part II, we describe working with students to write about ideas that they have vetted with one another: This happens in a short time frame as we shape and refine a definition, explanation, or diagram, and over a longer time frame as students provide peer review, write exams, and ultimately submit a final paper on a topic.

Of course, the divide between formative and summative writing is never that tidy: In producing more formal texts to disseminate ideas, we further shape and refine the original ideas. And much of the early work that happens as students develop ideas feeds into the more formal descriptions written later on in the process. (Ideas in science are in draft form for a very long time!) So this distinction we are making is a fuzzy one. As colleague and coauthor Kim Jaxon is fond of saying, "As the ideas become clear, the writing becomes clear." And so developing clear, succinct writing is always a process of refining and clarifying ideas.

That said, we find that there are moments when we have some "foothold" ideas: ideas everyone in the class is using, and using in similar ways. Or a group has conducted a series of experiments to refine an idea, has shared the analysis, and has been using that idea somewhat unproblematically to answer new questions. We may feel that a puzzling phenomenon has been explained. It may seem tedious to students to revisit these conclusions: A day or more of investigation and discussion has led us to an idea, and returning to an idea that feels "settled" is a challenge. However, after we begin the process of constructing formal, conclusive statements with attention to detail, students become aware of gaps in their explanations or discrepancies between one another's ideas, and developing clear, concise statements becomes a meaningful and challenging task. It may even spark questions that take us back into our inquiry, and we iterate: returning to and refining our definitions, explanations, and representations over time. Frequently what we ultimately are working on can be considered terminology: primary color, focus, orbit, "still"—even invented terms: Seurat spot, "second," hot sauce theory.

In many science courses, students are taught definitions (e.g., velocity is a vector quantity that indicates the rate of change in position), and they interpret and employ these in developing arguments or solving problems. These definitions are presented as incontrovertible, if not altogether obvious, as if they were the launching point of inquiry rather than a hard-fought, hard-won product of scientific inquiry. And while much scientific research does proceed from well-defined terms (and knowing how to read, interpret, and use those terms is critical), this is hardly true for all research. Defining terms (e.g., species, acceleration, life, acid, planet, particle, etc.) is a challenging and iterative process that underlies a great deal of scientific inquiry. As Bazerman (1988) notes,

> In Bacon's day the word *acid* meant only sour-tasting; then it came to mean a sour-tasting substance; then, a substance which reddens litmus; then, a compound that dissociates in aqueous solution to produce hydrogen ions; then, a compound or ion that can give protons to other substances; and most recently, a molecule or ion that can combine with another by forming a covalent bond with two electrons of the other. . . . The tasting and taster vanish as the structure emerges. (p. 164)

That is, as progress was made on understanding acids, the definition changed not only in its precision but also in its nature, shifting from a subjective experience to an operational definition to a theoretical account. For Galileo, determining whether he would define acceleration as change in speed per unit *time* or change in speed per unit *distance* was of considerable importance; the most useful choice is not at all obvious prima facie.

In class, these activities generally take 20–30 minutes of time. Often they are impromptu, as when the instructor notices that an idea is being used in different ways and pauses to have students clarify it, or notices that students would benefit from a clear definition. Small groups might work to construct their own definition before the whole class reconvenes to share, develop, and refine a consensus definition. Or each group might generate a summary paragraph to clarify an idea the students have been developing, and then share the paragraphs with the whole class.

Significantly, and unlike the "lab report" that often concludes the end of a unit, this summative work does not signify the end of our inquiry. We do not construct these ideas solely as a performance or to display to the instructor that the correct ideas have been learned. Instead, these ideas are clarified because such clarity is *useful*. Students are willing to do the labor because it is information that will have a use beyond assessment; the clarity will have a role in their ongoing work and in making their ideas understood and used. The definitions are used and matter for developing ideas.

Returning to our theme of similarities between coursework and scientific work, a paper in science is judged, in part, by the number of citations it receives after publication. Ideas have merit to the degree that they are *useful*. Even papers that are seen as fully concluding a line of inquiry are employed extensively in future work.

This is not to say that a clear idea is not an end in itself. It is. It can be immensely satisfying to arrive at a clear, concise statement of our ideas. This is often the primary outcome of scientific research. This work—in which students create clear consensus statements—refines their ideas and becomes a tangible product of their inquiry.

TAKEN UP

We generate several different kinds of summative statements in class: brief abstract-like summaries, a consensus diagram, and definitions, to name a few. Below we present two examples from our course where students work together to define a term; our work on other brief consensus statements follows a similar pattern. In the first example, students have been using one student's invented term—a Seurat spot—extensively over the past weeks, and the class works to clearly define this term. We begin with the Seurat spot example because this term—to our knowledge—has no equivalent term in physical optics. There is, then, no "right" way to define a Seurat spot, and yet, in trying to construct their more precise definition, they begin to ask significant scientific questions with bearing on canonical physics ideas.

After they construct that definition, it is a concept that has continuing relevance. In particular, the Seurat spot is taken up later when defining "focus," which students ultimately describe as a "Seurat spot reunification point" (SSRP). Both of these definitions are described below.

Seurat Spot

We often begin our inquiry into light by using pinhole theaters, as described in Chapter 4. These—like any pinhole camera—use a small hole to restrict light from reaching a screen and project an upside-down image on the screen. To develop a model of how this happens, students must develop models of how light reflects from objects, passes through the aperture, and projects onto the screen. For many students, this forces them to reconsider their (usually tacit) ideas about images. Implicitly, many of our students think of images as complete "things," rather than a composite built up from individual light rays. One student, Cassandra, has an "aha" moment and animatedly describes these points that comprise an image as "Seurat spots"—sharing a picture from the pointillism painter Georges Seurat with the class.

A few days after Cassandra's introduction of the Seurat spot, another student, Adam, describes the reflection of light off surfaces as "half disco balls," reflecting light in every direction. He constructs a physical model to show the class what he means—a ping pong ball with strands of spaghetti representing the light rays.

These ideas—the Seurat spot, spaghetti rays, and the half disco ball—had been pivotal in our first unit on light. Now, in our second unit, on the eye, the ideas were being used as students worked to describe what a lens does to light. As they worked, we recognized that there were variations in the way they were using the terms: Is a Seurat spot essentially the termination of a light ray (as represented by spaghetti)? Is it a point on an image, or on the object that casts an image, or both? Do many rays originate from one Seurat spot, or does each spot reflect just one ray?

The instructor, Leslie, notices this and suggests: "I think we're all using this in slightly different ways—so I want us to clarify. Can you finish the sentence: A Seurat spot is . . . ?" As the students discuss their ideas, the instructor takes notes on the board:

Adam: [to Cassandra] So you're thinking that all the spaghetti strands are a Seurat spot? Right?

Cassandra: I—I'm more along with Amy where it's like the collection of the rays are each a Seurat spot that make an image.

Adam: Yeah. So every piece of spaghetti on that ping pong ball is a Seurat spot.

Amy: Well it's a light ray.

Cassandra: It's a light ray, but, when it recreates the image, it's a Seurat spot.

Adam: So you're saying that a Seurat spot sends a ray of light?

Kait: But you have multiple light rays coming off from the same Seurat spot.

Amy: No—coming off of the image in the disco ball-ish pattern, going off of everywhere, and they're coming off. But they're not actually coming off of like, the exact *same* spot.

Adam's representation describes light reflecting off a surface in many directions, but does not specify whether the multiple spaghetti strands represent what happens to one ray, or to a collection of rays. Similarly, it is not clear whether the half disco ball is representing a single point (which reflects in all directions as if it were a disco ball) or whether it represents an area where each point reflects an incoming ray in a different direction. Cassandra, who had developed the Seurat spot idea in the context of a pinhole camera (in which one can assume that, essentially, one ray is responsible for one spot), had not considered whether multiple rays might construct a single Seurat spot. All these ambiguities are brought to the fore in discussing a definition for the Seurat spot.

When Kait suggests, above, that multiple rays come from one Seurat spot, Amy disagrees. Amy explains her position, noting that a light ray (which generated the spot) cannot break into multiple rays:

> *Amy:* Because a light ray can't, like, break. So it's like two light rays that are really, really close to each other and they can go off, or two light rays come in really, really close to each other can go toward the same place.
> *Trevor:* That's the problem, though—because it's one single spot *and* it's got light rays going in all those directions. It's not a bunch of little spots sending out a bunch of light rays.
> *Kait:* It's the same exact spot.
> *Trevor:* Yes.
> *Amy:* But light doesn't do that.

(The ideas voiced here were justified earlier: Trevor justifies his claim by noting that images do not look "patchy," and Amy's position that a ray does not "break" is a commitment to an idea she learned previously about light. That is, these claims are based on evidence or text, although these justifications are not raised in the moment.)

This distinction, it turns out, is not easily resolved. And so, in trying to clarify her position, Amy and Trevor hit on a significant question about the nature of light: Does one incoming ray "break" into many outgoing rays after striking an object? Below, Adam re-imagines what a light ray represents, suggesting that the ray in their representations is describing a collection of tiny, smaller "rays"—each of which can reflect in a different direction. The ray does not split or break; it is composed of smaller pieces, each of which travels in random directions after striking the surface:

> *Trevor:* It's not splitting. It's an individual light ray that's heading in any direction.
> *Adam:* Think of a light ray as only like a centimeter long and there's just thousands of little light rays hitting that one spot to go in random directions.

It's not uncommon in our course for a few students to engage in a discussion or debate that begins to leave others behind. The instructor pauses to summarize the two ideas that are behind the initial debate, sketching on the board. And Trevor clarifies the third idea:

> *Leslie:* So is it like light comes in and hits this disco ball and it breaks into lots of little rays?—So that's the shattering idea. Or is it like

one ray comes in and [drawing it so that it hits just one small part of the disco ball] heads that way, one ray [drawing another ray that hits nearby] comes in and heads that way, one ray comes in and heads that way?

Trevor: I think the problem is that it's a constant flow of light and *not* only one single ray—it's like a river dumping onto this thing. So you can keep sending light rays out in all directions.

In trying to define a Seurat spot, Trevor (and, earlier, Adam) has added nuance to our understanding of what a "ray" in our diagram can represent. We have reason to believe that light travels in straight lines, and we have reason to believe that it scatters out in all directions from a surface; but the line itself—the "light ray" we are drawing—need not represent "light" so much as a path along which smaller segments of light travel. Kait follows up by echoing Trevor's idea and explaining that these small rays are too small to draw individually:

Kait: So we're just assuming that the thing hitting it is so small that we're just saying for ease of drawing we're just ignoring that they're coming in at different times.

Tom: You would have to break it down into a time period so infinitesimally small we can't even imagine it. It would have to be like a trillionth of a billionth of a millionth of a second.

Amy, hearing this, tosses a question back to the class—a point of consensus that can ignore the question of the size of the Seurat spot:

Amy: Does it make sense to you guys—or everyone—that an object has light hitting it, bounces off in light rays, and forms a Seurat spot that makes up an image?

Adam: It has to go through a pinpoint or a pinhole or a lens first.

Amy: Yes, yes.

Adam: But yeah.

Amy: If that makes sense . . .

Esmerelda: It's not just one Seurat spot that makes it—it's a whole bunch that came off that object.

Amy: A whole bunch of Seurat spots to make up the image of the object . . .

Adam: It *becomes* a Seurat spot once it hits an object.

Leslie writes on the board, "Every ray of light (spaghetti) creates a Seurat spot when it hits an object." The class agrees on this statement (clarifying that this object may be a retina).

This careful attention to a definition is characteristic of academic writing, particularly in science. Moreover, it isn't something that is done on the first day—we do not start our inquiry with precise definitions, nor do we start our inquiry with well-articulated "testable questions." Like the construction of all scientific facts (e.g., Latour & Woolgar, 1979), precision in both definitions and questions necessarily emerges over time.

For readers who are familiar with the canonical topics of geometric optics, it may be surprising to see a class spend so much time on an idea—the Seurat spot—that, while consistent with those topics, has no canonical equivalent. However, we can see that the attention to this idea generates questions that are central to optics and light. Moreover, the idea of the Seurat spot is present in optics, but implicitly. And so, in defining the Seurat spot, students are describing the set of assumptions behind laws and rules that will come later: how images are generated and the role that lenses play.

To be clear, not all questions about Seurat spots are resolved in the definition: not in the discussion transcribed above, not in future iterations of the definition, and not during the course of the semester. The question of the "size" of a Seurat spot, for example, is intriguing: Should this be a fundamental limit of nature—the size of a photon? Or should it, instead, be dictated by the size of an aperture that is used to construct an image? Or should we think of the size of a Seurat spot as defined by the biologically related structures of our retinas and the limits they impose on resolution of an image? The students do not address these questions in this course. They focus, instead, on a more pragmatic description that avoids a commitment to size.

As a definition, it could use more work: We are still vague on the nature of a "ray," for example, and we probably should use the word *light* to clarify the nature of the ray. It would sound more like a definition if it read, "A Seurat spot is generated by a ray when it hits an object." In addition, we have not yet studied color, and so students will likely revise this when learning that some rays are reflected and some are absorbed (in particular, the rays that are reflected are those that generate a Seurat spot on a screen; those that are absorbed create a Seurat spot on the retina). Nonetheless, this succinct definition encapsulates a significant idea, is agreed upon by the class, and enables us to use the term consistently as we move forward.

Focus

With the same group of students, we moved from describing pinhole theaters and light to understanding how the eye works. One of the more perplexing structures in the eye is the lens; while it seems clear that a lens helps to "focus" light, how that happens is not clear. Groups approach this problem in a range of ways: Some, using a Maglite, notice that light spreads out as it leaves the light bulb, passes through the lens, and then comes back together at a point before fanning out again. Other students have noticed that light from two lasers—parallel to each other before entering the lens—will cross after the lens at a point they call the "flipping point." A third group has examined the lens' ability to project an image on paper. All groups have been trying to integrate these ideas with the eye and our ability to see clear, focused images.

Now we ask students to clarify what they call the "focal point," using their whiteboards to write out their explanation. (To scientists, the term *focal point* is the point at which parallel rays meet; students are not talking about that point but, instead, what usually is described as the image plane of a lens.) Students are brought back to their lab groups to share their ideas and develop a clear description of what happens when a lens creates a focused image.

(For those unfamiliar with how lenses focus light, the basic idea is that rays, which usually travel in straight lines, bend when they enter and exit a lens. The shape of a lens is such that rays that leave one point are bent in such a way as to all be brought back together at one point.)

The definitions on the six whiteboards are as follows:

Group 1: "The focal point is the point behind the lens (on the retina for a clear image) where all the light rays from one minute Seurat spot carrying one single piece of information meet up again to form one single Seurat spot."

Group 2: "The lens changes shape to focus and direct light *onto* the retina into a single point. That single point is the focal point."

Group 3: "Focal point—a point of convergence. As for our eye, it is the point where all of the light rays from a single Seurat spot are angled back to a single Seurat spot, this focal point being on our retina when the image is clear." [See Figure 7.1]

Figure 7.1. Student Diagram for Definition #3: Rays from a Seurat Spot Pass Through the Lens and Come Together at the Focal Point Called a Seurat Spot Reunification Point

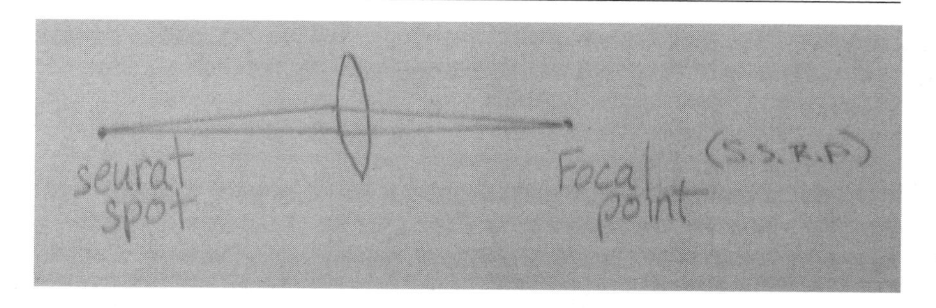

Figure 7.2. Student Diagram for Definition #4: Comparing a Focused Image on the Retina ("4 Stamps on Same Spot") with an Unfocused Image on the Retina ("4 Separate Stamps")

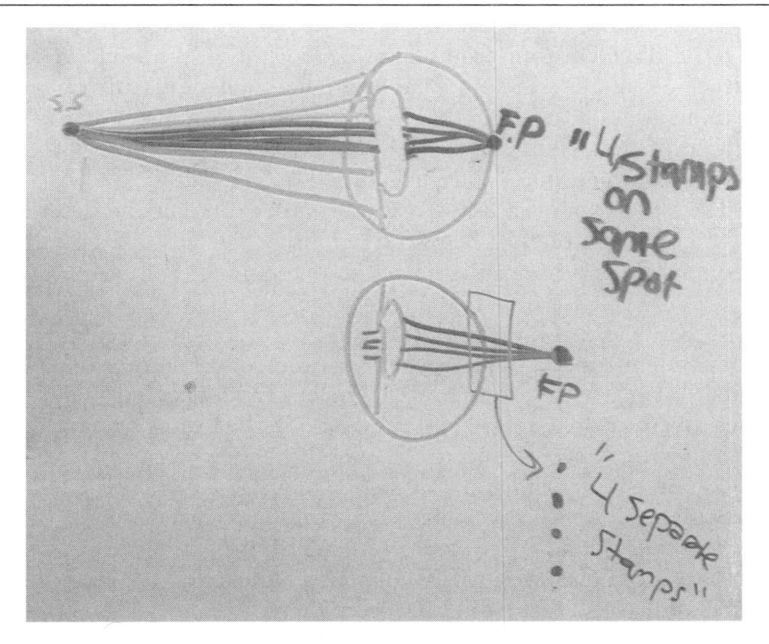

Group 4: "Focal point is where all light rays that enter through the pupil into the lens that are coming from one Seurat spot meet together." [See Figure 7.2]
Group 5: "A focal point is where all the light rays that pass through a lens from a Seurat spot meet at a single point."
Group 6: "We don't think the focal point is a point at all. We think it's an area on the retina where our Seurat image is in focus. This could be thought of as the point in the light rays when they are in focus. We think there is a point in the vitreous humor where all the light rays cross (flipping point) then continue to create an image on the retina."

Some things to notice about these definitions:

- These are not initial ideas about how a lens will focus light, but represent, in many cases, commitments students have made based on their recent work.
- The definitions vary and reflect questions that the individual groups have attended to. For example, the first definition is careful to note

that the focal point need not be on the retina, but if it isn't, then the image will not be clear to the viewer; that group also thinks of the multiple rays from one Seurat spot as carrying "information." The second group attends to the shape of the lens, as the students have looked at rounded and flat lenses, but gives less attention to the Seurat spot. And the final group highlights the truly remarkable thing about a focal point: A focused image is, in fact, an area—not a point at all—and so we need to be concerned not just with a focused point, but a collection of focused points.

- The Seurat spot is a useful term for understanding and describing what the lens does: It helps to clarify that the lens takes rays from one point and brings them back to one point.

As students work on their whiteboards, the instructor circulates around the room and imagines a trajectory of ideas and conversations. In particular, I imagine starting with Group 6: Its members are concerned that an image occupies an entire area and is not a "point" at all. I want to have the class understand their concerns and try to take those into account as we discuss how the "points" that other groups define might connect with a full image. (In particular, Group 2's text suggests that all light, not just light from one spot, is directed to one point on the retina.) After hearing from Group 6, I plan to hear from Group 2, whose members indicate that focus is achieved by a lens' ability to direct light, and, from there, have the class consider the more precise descriptions. For Group 3, SSRP strikes me as a better name than "focal point"; the name itself is descriptive of what has to be done to achieve focus, and we imagine that Group 6 will find SSRP a clear term and definition—one that avoids the concerns they have with "focus." And so I want to call attention to this early and use that term in our conversations.

Groups 2 and 3 both describe the importance of this point being on the retina if the eye is to perceive a clear image. As for Group 1's idea, I love the reference to "information": It suggests that an image is a precise ordering of information, and this idea connects with the concerns raised in Group 6's whiteboard—that "focus" should not be defined by a point, but an image. Group 4 brings up the metaphor of an image as a "stamp," and an unfocused image is a series of overlapping stamps. I want the group to discuss this idea with the class more broadly, as I think it helps to connect ideas about points and images: By considering a series of stamps overlapping precisely, we can examine the connections between a focused point and a focused image.

This attention to ideas and orchestrating the conversation to draw out important themes is a critical activity for the instructor in an inquiry classroom. From the outside, it can seem as if the instructor simply is facilitating the student-led conversation; to the instructor, however, there is a great deal of orchestration happening. The ensuing conversation is never quite as orderly as imagined, and the role of the instructor throughout is to help connect these ideas, table some ideas for later, and drive the conversation to help reach consensus when possible:

> *Leslie:* So I want to start with this group [Group 6] because I want us to consider their objections to the idea of a "focal point"—Amy, can you say a little bit about what you've written?

And then:

> *Leslie:* Okay—so let's just keep that in the back of your mind—I think a lot of the boards actually have something to say about that—but for now, I want to move on to Andrew—your group's board . . .

Later:

> *Leslie:* So on your board [to Group 3] you called this a Seurat spot reunification point, and I guess what I want to know—and [to Group 6] I think this is what you want to know, too—is that point a clear image, or is it just a point?

The conversation highlighted several critical ideas: first, the idea that the "job" of a lens is to create a Seurat spot reunification point by bending the rays back together. Second, to create a complete, focused image, it must reunify all the points in the image in their appropriate locations. Finally, we clarified that this "reunification" must occur on the retina for an image to be seen in focus. We did not press to write one single, shared definition (in hindsight, this would have been a productive step), but instead the conversation reached a verbal consensus, and students continued their work.

In other chapters of this book we discuss how ideas and definitions from the class get taken up in assignments. Here, we just note the following description taken from Kaitlyn's final paper: a description that brings in ideas from the Seurat spot, disco ball, and focus to describe succinctly that a lens focuses a point on the retina, and why images are blurry when that focus is not on the retina:

> For each spot on an object (Seurat spot) an infinite amount of rays are being reflected in all directions (like a disco ball). All the rays that make it into our eye through our pupil from one Seurat spot need to be focused to one single spot on our retina. If the spot was not on our retina, one single Seurat spot would end up looking like a blurry larger dot. The overlapping of these blurry large dots would make our image look very blurry.

CHALLENGES

Why not simply provide students with definitions for critical terms? After all, scientists are not actively reconsidering what is meant by "focus" in geometric optics. And although constructing terminology is a scientific activity, so is facility in using accepted definitions. In our experience, it is through the careful, iterative construction of definitions that students come to understand the nature of scientific definitions. That is, when we engage students in the careful construction of a definition, they are more likely to understand that these ideas are complex, nuanced, socially constructed over time, and embedded in a much larger process of model building and experimental evidence. In this way, by gaining facility in constructing definitions, students are better prepared to interpret and use other scientific definitions productively.

However, because students are often unfamiliar with this process, it can be challenging for them to recognize the precision and care that go into constructing a definition, and in using that definition as we move forward. As instructors, it is our role to push for clarity and consensus—not imposed from outside, but as a central part of our inquiry and how we understand and refine our ideas together. That is, the clarity serves a genuine rhetorical purpose for our students. We seek out and attend to differences as a way of motivating our work on definitions. As students generate definitions, we look for places of dissent and call attention to them: "Both of you are saying that the focal point is where rays come together. But this definition stipulates that those rays left one Seurat spot, while that definition suggests that all rays come together at a focal point."

We often are asked how to prevent students from simply looking up definitions online. When using student-constructed terms, of course, this isn't an issue: The Seurat spot, as described above, is one example; other examples include "buoyant potential energy" (the energy stored when a floating object is submerged), "seconds" (a term for diffusely reflected rays), and "streakiness" (a measure of a baseball player's hitting "streak"). When students introduce ideas like those, we promote them, taking them up in homework and class discussions. For scientific terms, we find that if we introduce this challenge as part of the ongoing dialogue ("you're using this idea in slightly different ways . . ."), students tend to understand the challenge and take it up, justifying their claims by reference not to an outside authority but to ideas that are developing in the classroom. When they do bring in an Internet definition, we often don't have enough context to make sense of it; for example, a popular physics site describes focal point as: "For a thin double convex lens, refraction acts to focus all parallel rays to a point referred to as the principal focal point" (Nave, n.d.).

Finally, a challenge lies in deciding when this more summative work should happen. For science faculty, who are used to working on well-defined problems, or at least in well-defined paradigms, early inquiries can feel ill-defined. We sometimes feel as though students would benefit from clear terminology, and we are tempted to push for this early in students' investigations. However, as Elby (2009) notes, it may be "more productive not to converge on a definition until further empirical and theoretical progress points us toward the best way to 'cut up [nature] . . . along its natural joints'" (p. 139). It is important to give students the opportunity to explore a topic and generate ideas so that, eventually, they have some definitions to propose and defend that are consistent with their models and evidence.

FEEDBACK AND GRADING

As described in earlier chapters, we provide opportunities throughout the course for students to receive feedback from one another. The brevity of our definitions allows all students to share, critique, and refine the statements during a class period and reach clarity and consensus. Instructors provide a great deal of feedback through this process, as we help students refine and revise their definitions by highlighting differences and pushing for clarity.

And, as mentioned in prior chapters, this work is not graded in the moment. Instead, we will ask students to take up the ideas in homework, papers, and exams. They are held accountable for understanding and using those terms appropriately in graded work. And, again, we understand that definitions may change over time as new models are introduced and new data come to light. We allow for this flexibility in our class as well; student definitions, for example, of "primary color" frequently will change, even after several weeks of inquiry.

Take-Home Messages

- Producing formal, summative writing should happen throughout a unit, as ideas are vetted and established. This should not occur only as the final activity for a line of inquiry.
- Formal, summative writing does not have to be a lengthy paper. Careful, detailed attention to writing a definition, a succinct explanation, or a representation is something that can be tackled during a single class period, with attention paid to every word.

- This work can help make visible to students the "behind-the-scenes" negotiations that are critical in scientific work; constructing a scientific definition, for example, is neither straightforward nor wholly empirical. It is scientists who determine definitions of terms through ongoing negotiations. Engaging students in constructing definitions can help make the decisions and the social construction of knowledge more transparent, and reduce the appearance of science as a "rhetoric of conclusions."
- Students should recognize that precision is useful for the purposes of understanding one another and making further progress: If we cannot clearly understand ideas, we lack the transparency necessary to vet an idea, and without clarity, it is difficult to make productive use of an idea. Precision is something that faculty demand not because science uses precise vocabulary but instead because this precision is necessary to accomplish the inquiry being undertaken.
- The writing that students carefully construct should be used in class; just as the importance of a scientific paper is determined by the number of citations it has, students will recognize their ideas as important to the degree that these definitions, explanations, and representations are taken up in later work and that this later work is facilitated by the clarity of those ideas.
- It is often thought that precision is required in order to begin scientific inquiry: "Testable questions" and "precise definitions" are seen as characteristics of scientific practice and often required of students at the start of an investigation. This precision, however, is not always present at the start of our inquiry, nor can it be. Instead, precision is a product of ongoing work and can help shape later inquiries.
- Ideas—even clear, well-constructed, "summative" definitions—will change. This iterative defining is characteristic of science; as theoretical progress is made, our definitions change. Ideas should be accountable to the available evidence, consistent with the current models, and work toward coherence.
- Drafting a consensus statement means that you cannot assign a grade to any one student. These are collaborative efforts as the class works to clarify a particular idea. Solicit input from as many in the class as you can, and structure the conversation to incorporate a range of ideas. Provide feedback throughout the process and hold all students accountable to the idea and its use in their individual assignments.

CHAPTER 8

Final Papers

> Our [final papers] were far from the traditional lab report. I liked how we had the freedom to write and not be confined to strict guidelines of how a certain format had to be. . . . The written assignments were quite enjoyable because I was able to go over all the notes I had taken, and decide which, what, or how I was going to create the paper. —Johnny

> Sometimes it's hard to talk about it with people who are not in the class because they don't have the group background we've built together. —Ariana

In other chapters, we describe the ways in which scientists use texts that often are not considered "academic writing," and advocate for their place in the science classroom: lab notebooks, whiteboards, graphical representations, peer reviews, annotations of readings, and the joint construction of definitions. Here we turn toward texts that usually *are* what faculty think of when they think of "writing" assignments: the more lengthy expository paper, submitted to the instructor for a grade at the end of a unit, that ties together experimental evidence and scientific models.

Parallels between these assignments and scientific practices are obvious: The culmination of a productive line of inquiry, for a scientist, is usually the scientific paper. In these papers, findings are introduced that fill a gap in our understanding, answer an open question, launch a new line of inquiry, describe a novel technique, and so forth. In many writing assignments for science courses, students are asked to generate similar writing, often employing the structure of a journal paper (with introduction, methods, data, and analysis) to inform their own writing. A common school-equivalent of a journal-like composition is the lab report. (A brief Google search of "college science lab report" leads to numerous campus help pages and advice for students; the lab report is clearly alive and well in the university.)

However, an emphasis on the structure of the lab report can ignore a more significant feature of the scientific paper. As mathematician Halmos (1973) notes, in an essay on writing,

> It might seem unnecessary to insist that in order to say something well you must have something to say, but it's no joke. Much bad writing, mathematical and

> otherwise, is caused by a violation of that first principle. . . . To have something to say is by far the most important ingredient of good exposition—so much so that if the idea is important enough, the work has a chance to be immortal even if it is confusingly misorganized and awkwardly expressed. (pp. 20–21)

While it can be hard to have "something to say" in reference to scientific phenomena, it is harder still to organize and write a meaningful paper when you have nothing you really want to say. The structures designed to help with the exposition of an idea—a five-paragraph essay or the headings of the typical lab report—become a MadLibs of science writing as students, without a clear idea to share, seek the "right" words to fill in the template. Rhetorical structures, which may be useful in organizing ideas and helping us find the holes in our arguments, can do only so much in helping us have an argument to make in the first place. If students "cannot" write a topic sentence, it is likely that they have no topic sentence. If they do well on all but the "analysis" section of a lab report—and why report on a lab if not for the analysis?—then it is likely that they have no real analysis to share. These structures can help highlight that, but all the writing instruction, rubrics, and templates cannot give students something to say.

It is, therefore, critical that students *have something to say*—a hard-won idea that they are proud of, a unique insight that they have developed, a representation or a piece of evidence or a way of phrasing an idea, or even a question that their investigations have helped them articulate. One aim of our course is to help students have scientific ideas they want to share. That, in a nutshell, can be considered the underlying goal of lab notebooks, whiteboards, class discussions, definitions, and reading annotations. And the purpose of the final writing assignments is to give students a place where they can say it.

Of course, traditional courses and lab reports do provide this for some students, some of the time: We certainly recall high school and undergraduate courses where we felt as though there was something significant that we had figured out and wanted to write about, and we can remember assignments that provided that opportunity (some 25 years after the fact). One author distinctly remembers writing a paper regarding what was meant by "equal and opposite," and even a seemingly routine proof for the field inside a spherical shell was a source of great pride when rederived from scratch. And, of course, scientists employ routine structures in their writing, with many journals providing style guides to authors. But for many students, particularly nonmajors, writing assignments fail to provide a space where they share their own ideas, and the structures we provide are distractions from that fact (Wiley, 2000). We advocate, therefore, for careful attention to developing students' ideas and generating writing assignments that value those ideas. In this way, the assignments function like scientists' journal papers, although they frequently lack the structures that journal papers have.

In addition, students not only need to have something to say, but *they should have someone to say it to*. Halmos, in the essay referenced above, notes: "The second principle of good writing is to write for someone" (p. 22). In many cases, the literature review that starts a scientific paper is as much about defining the "someone" as it is about situating the work in a particular tradition or set of questions. The two go so hand-in-hand that "having a scientific idea" practically requires that one is engaged in a scientific conversation—that is, scientific ideas are socially constructed and will have a particular scientific audience in mind. Although scientists (and many others) create popularized accounts for nonscientists, the science itself is generated in dialogue with other scientists, building on open questions, challenging settled questions, or providing a unique analysis, a new technique for a defined field, or data to use in selecting between competing ideas. The intertextuality of scientific writing is, in large part, what makes it so hard to read: These texts hardly stand on their own, but require a knowledge of the field, the open questions, the competing ideas, the peculiar definitions and mathematical constructs, in order to understand their claims (Sharma & Anderson, 2009).

Moreover, there has been a lot of attention to the value of a "real" audience for writing assignments (Ede & Lunsford, 1984; Weiser, Fehler, & González, 2009). This often is construed to mean "someone outside of the classroom," as a way of ensuring (to the degree possible) that the writing is not a school-based performance, but is situated in a genuine rhetorical situation. But students cannot, in general, address open questions in scientific fields in one semester (citizen science and other data collection projects notwithstanding). The point of a dissertation—years in the making—is to show that you have something to say and a community of scientists to say it to (some of whom must agree that the thing you said was worth saying). We certainly are not advocating that undergraduate students generate new science for a scientific audience.

And so our course must create this community among ourselves. In our courses, the audience, although wholly within the classroom, is nonetheless "real": Students read one another's work because they are interested in it, want to find out about it, and want to weigh in with their own thoughts. Again, this is not because our students are inherently interested in color, light, or any of the other topics—this interest, both in the topic and in their peers' ideas, is cultivated in the classroom. Through our construction of ideas, the feedback, and iterative work that we do, students are vested in one another's ideas, findings, and claims. The arguments they make are designed, over time, for one another. And much of the peer review work outlined in earlier chapters is how this community develops.

TAKEN UP

Before proceeding, let us quickly describe what we mean by "something scientific to say." What does this look like in the context of nonscience majors in an undergraduate course? There is a concern when you move away from the textbook, clear laboratory procedures, and defined content outcomes, that student ideas will not be aligned with disciplinary knowledge in any meaningful sense, and research on "open inquiry" bears this out (Kirschner, Sweller, & Clark, 2006). To be clear: These classes are not purely "open" inquiry, but instead highly structured in the ways outlined in earlier chapters. Below are two examples of the kinds of things students "have to say" in our courses. Some things to note are that they strike us as deeply scientific and insightful: The ideas are tied to arguments, questions, and data. But they are not, in general, wholesale theories, nor do they necessarily represent weeks of development.

1. In one class, students were puzzled by the "fuzzy edge" of shadows as they shone a light down a long, black tube. One student proposed that this edge was due to rays being reflected from the tube; another articulated that they were "fuzzy" in appearance because some of the light was absorbed when it was reflected. In a later conversation, someone questioned whether light reflecting off a mirror should be considered a "fuzzy" ray (rays they called "seconds"). Another student suggested that mirrors reflect all the light, so these are not "seconds." We suggested that this could easily be measured, but another student suggested this wasn't necessary: "Mirrors must absorb light. Metal slides get hot."
2. As we studied the eye, this question came up: Why do you have to look right at a word to read it? You don't have to "direct" your ears to hear a word. Two possible answers were generated: (1) images are truly in focus only at that point directly behind our lens—the rest is blurry, and (2) we have better resolution in the center of our vision—perhaps the receptors in our eye are closer together there—so reading with our peripheral vision is like trying to write in 12-point font with a crayon. In discussing whether we could even tell (with our own eyes) the difference between blurry and low resolution, a student noted: "Well, you have to look directly at the sun for it to hurt. So the sun must be out of focus in your peripheral vision."

We could recount similar insights from almost every day of the course: The student who decided that the eye must have receptors that mimic the colors on a cellphone screen, and that her color-blind lab mate must be missing one of those colors. The student who challenged our model of diffuse reflection by asking why objects don't appear to "glow." A disconnect between our theory that a more rounded lens should focus light, and the sense that our eyes strain to focus, suggesting the muscles are pulling the lens taut to focus on nearby objects. When we say that students are authors of their own scientific ideas, and that they are, indeed, disciplinary, these are the kinds of ideas we are talking about. Students are engaged in developing models of phenomena and coordinating those models with evidence, both "scientific" evidence and everyday observations. They are collaborative, building on or challenging other students' questions and models. It is these ideas that we want to honor with our writing assignments, giving students a place to share those insights.

Each 5-week unit ends with a formal writing assignment. These are lengthy papers (5–10 pages, in general) that tie together major ideas from the unit. We have used two types of final writing assignments in our courses: (1) assignments, which we write a day or two before they are handed out, that are tailored to each lab group's specific observations, data, and explanations, and (2) a common assignment, determined in advance, based on students' ideas. Both are described below, with examples of how they are connected to students' own ideas.

Assignments Tailored to Students' Ideas

Below are three of the assignments from the first unit of the semester. The topic was color (beginning with the question, Is every color in the rainbow?), and students had developed questions around lights, paints, pigment, and screen displays. In particular, students had a model wherein the primary colors of paint were cyan, magenta, and yellow—all other painted colors were assumed to be a blend of these—and red, green, and blue were the primary colors of light, because of their relationship to those primary paint colors (namely, that cyan absorbs red, and so on).

One group had developed a technique for investigating the appearance of printed colors under colored lights. Students printed a "rainbow" spectrum, placed it in a box, and used colored bulbs to illuminate the paper, which they then photographed with their cellphones and shared on our course website. This experiment was designed to examine the notion that red paint reflects red light, but not other colors. Their data complicate this somewhat, in that cyan paint seems to reflect more blue light than blue paint reflects (the cyan shines brightly under blue light, while the blue paint looks quite dim), but they have not attended to that yet.

For their assignment, we asked them to explain why. In doing so, they would need to articulate the models they (and the rest of the class) had developed and use those to explain this observation. As the models fall short, we anticipated that they would need to reconsider their model for either blue light (this bulb is not likely a primary shade of blue) or cyan pigment.

> ***Assignment 1:*** Explain why, under blue light, the cyan stripe practically blends in with the white background but the blue stripe does not. Shouldn't blue light on blue ink appear to be the same color as blue light on white paper (at least, that's what Estefan's group argues)?

Another group used the above technique to examine paints, which seemed to behave differently from printer inks. (Paints are not always made by mixing primary colors. That is, green might be made from green pigment and not from a mixture of yellow and blue pigments. Students do not know this.) They painted stripes of color on paper: blue, green, yellow, orange, and red. To generate yellow light, they used red and green bulbs. When they shone the yellow light on the colored stripes, the yellow paint, as expected, practically blended in with the white paper since both reflected all of the yellow light. More surprising was that the green paint appeared orange and the blue paint appeared to be almost the same shade as the red paint. Again, to explain this they needed to draw on the existing theories and consider implications for what this evidence told them about the nature of the blue, green, yellow, orange, and red paint. The assignment included a photo from their investigation, and a range of data was available on the spectrum of the red and green bulbs that students might draw from to justify their responses.

> ***Assignment 2:*** You tried to shine both red *and* green light in your box at the same time. Explain at least two of the colors and why they appear this color under red + green light. This is a tough question, and your job is not to get the "right" explanation, but to clearly explain your own ideas.

We knew that a third group's observations on fluorescent lights and incandescent lights had not been considered in terms of the implications for seeing particular colors. That is, we had discussed blue light and the way it interacts, say, with red pigment. But we had treated all "white" light the same, as if the white fluorescent light, with its distinct bands of color, was equivalent to the incandescent bulb with its complete spectrum of colors. We hoped students might recognize that a pigment (say, something between cyan and green) that did not have an equivalent wavelength present in the fluorescent spectrum would be much dimmer when viewed under the fluorescent light. To justify this idea, they would need to use theories we had developed and build on those to consider the particular shades that are not present in fluorescent light.

Assignment 3: We know that a green pigment under red light looks black. Are there any pigments that would look different under fluorescent light than they do under sunlight? Below are pictures you took of sunlight (left) and a fluorescent light (right). This is a tough question, and your job is not to get the "right" answer, but to clearly explain your own ideas.

Common Assignments

We don't always write assignments that are specific to a particular lab group, or even for a particular semester. For example, we assign an "eye paper" at the end of our unit on the eye that simply asks students to "explain how the eye works." This may seem like a vague and easily "Google-able" assignment. However, the sequence of homework leading up to this final assignment, together with our work in class on constructing ideas and definitions, leads to a highly personal account of each student's understanding of how the eye works.

In particular, our homework assignment at the start of the unit on the eye always includes a question along the lines of, "Why do we need an eye at all—what would happen if we had a retina on our skin?" In this way, students begin the unit by recognizing some of the problems that the structures of the eye must solve. Having articulated these problems themselves—that is, clearly defined them, so that "blurriness" and "too bright" are not just vague ideas, but carefully constructed terms and models—they are better positioned to describe the functions of the iris and lens. Later, we ask a version of the question, "What happens when the retina gets an iris?" Here students, who are familiar with pinhole cameras, begin to consider the scale of the eye and the degree of blurriness that is present even with a small pupil. The arguments they make become more sophisticated as they draw on experiments and observations they have made, along with models they have constructed. And finally, we ask about what happens when a lens is placed behind the pupil. Models of focus and how the lens achieves focus at a range of distances must be addressed.

By the time students are writing about the eye, then, they are not rewriting a set of lecture notes or a book report that describes their interpretation of text. Nor are they filling out a lab report of hypothesis, methods, data, and interpretation. Instead, they are highlighting problems that others generated ("the periphery, however, is not likely to be in focus . . ."), with vocabulary the class has constructed (the Seurat spot reunification point), referring to experiments they have done ("because we saw that the lens inverts the image . . ."), answering questions that were relevant to their class ("these rays are made up of smaller rays of light"), and drawing on readings that they have critiqued ("Berkeley describes this as 'confused vision' . . ."). Every semester, these ideas are tackled in different ways, with students attending to slightly different questions and building on slightly different models. Some semesters, students are vexed by the question of how the eye changes focus; other semesters, as noted above, they work to understand why we see more detail in the center of our vision. That is, even though these assignments are not tailored to student ideas, they are embedded in a class that is, and the expectation is that these assignments will give students an opportunity to share their ideas.

These papers, then, lack some of the conventions of "academic writing" that many have come to expect in courses that introduce students to science writing. For example, we do not insist on a particular structure or require a particular format for citations. It is not essential that students embed figures and title them in a particular way. We sacrifice this fidelity to scientific writing in support of writing that offers students an opportunity to share their own ideas. We do not think the two are mutually exclusive; we can imagine a course such as ours in which, after developing these ideas, we would work to establish style and formatting conventions for our writing. However, given our student population and the time constraints, this is not our focus.

CHALLENGES

When students begin to write, they often fall back on idiosyncratic conventions and expectations about what a "paper" should look like. For example, they may start by saying something like: "The eye is a very important organ in our body . . ." or, "Many people wonder how the eye works. This paper will explain in three steps how the eye works." These rhetorical moves are not necessarily problematic, but they do suggest that students are framing this paper as a more formulaic essay or lab report, rather than an extension of the kinds of writing we have been doing all along.

When we notice students doing this, we usually ask them why. Is it a familiar structure and they find it helpful, or, instead, do they think that this is the approach we were expecting? When students find the structures useful, then we encourage them to use those structures. More often than not, however, they find it difficult to use these structures to explain their ideas. So much of our work is not a progression from hypothesis to test and analysis, and recasting it as such (while common in the sciences) is not necessarily the most meaningful approach for students to take. Those trying to write "five-paragraph essays" generally find that their claims do not have three supporting paragraphs. And so we ask them to map out their own questions that they resolved, and the explanations and justifications they generated, and then use those to construct their paper. Or we may suggest they look back over the readings we have used in class and use those as models to start or structure a paper.

Of course, often it is only after this paper has been submitted that we notice that students have framed it as a "lab report" or essay. In those cases, we grade, as noted below, on the content of the paper.

Finally, as noted in the Introduction, this is not a book that emphasizes grammar and other conventions for writing. And we also noted earlier that "as the ideas become clear, the writing becomes clear." But that is not to say that our students all write beautifully constructed, clear sentences and well-organized papers. We read our students' final papers knowing that they are newcomers to our field; they are trying out the ideas in science, and as they try out challenging ideas, the sentence structures may not be tidy (Bartholomae, 1985; Lunsford & Lunsford, 2008). We attend to errors on final drafts, but only when an error is competing with an understanding of the ideas the writer is trying to convey.

We're often asked whether students ever "Google" an answer. There is no shortage of explanations online for vision, for example. However, by the time we have reached the end of a unit, as we hope the earlier chapters illustrate, students have a rich set of ideas, representations, and experiments to draw from so that a "Googled" answer would stand out—both to their peers, who will review their writing, and to the instructors as we read the assignment. More important, incorporating "Googled" answers with their ideas, representations, and experiments is not trivial; those who could do this effectively would, essentially, not be copying ideas from an authority, but would have to interpret those in light of their own ideas. That is to say, using external resources has not been an issue for our final papers.

FEEDBACK AND GRADING

By the end of a unit, students should have received feedback on their ideas for 5 weeks: in lab groups, in whole-class discussions, via whiteboards, in gallery walks, in silent science, on homework—almost all of our interactions are structured to provide feedback from their peers and from faculty. Because our exams explicitly draw on these ideas, there have been numerous opportunities for feedback on students' ideas over the past weeks.

We generally provide some brief class time to garner additional peer review. For example, we might ask students to bring in one paragraph, one diagram, or other short excerpt they have been working on to share with their lab group and receive feedback. We also ask them to establish routines for getting feedback on drafts outside of class, such as setting up a time to meet with their lab group, or ways to share documents online with timelines and guidelines for that feedback. We often ask student writers to write memos to their readers. A memo gives the reader some context for the goals the writer has for the draft, shares what the writer likes about the draft, and asks for specific feedback the writer would like from the reviewer (see Jaxon, 2006, for more on setting up productive peer feedback). The reviewer uses this memo and considers the writer's ideas and requests as part of the feedback.

As with homework, when we approach grading we first read through the assignments and write a brief letter to each student that comments on what students are doing well and what we appreciate about their writing, and also includes any questions we have about their claims and inferences. We hold them accountable to the ideas generated in class: common terminology, models, data, and evidence.

After reading and responding to the writing, we cluster the papers of similar quality together. Generally, a few stand out as thorough and detailed, with thoughtful use of diagrams, clear references to evidence, and careful engagement with our established ideas as the students develop responses to new questions. Others will have similar strengths, but perhaps the diagrams are a bit imprecise, or the question that we asked the students to consider is addressed but without careful attention to the range of ideas and evidence available to them. Still others will offer more cursory explanations, not draw on our established ideas correctly, or have representations that are vague and confusing. As we sort through the papers and group them, we assign grades, on a 10-point scale. In general, although this varies, we may have two to three papers to which we assign a 10, and we'll have two to three that are a 6, with the rest falling in between.

Take-Home Messages

- To write well, students must have something to say. Prior to being given a summative, high-stakes writing assignment, students should have ample opportunity to develop their own ideas about the topic.
- An emphasis on students' own ideas is not equivalent to students producing opinion pieces or naive, "science fair" accounts: Allowing students to have their own ideas does not mean we abandon rigor. The prior chapters, in which students develop these ideas, provide examples of how we structure assignments and feedback so that this is not "anything goes."
- To write well, students need to write to someone. In our courses, this "someone" is (usually) their classmates and faculty. Prior to the summative, high-stakes writing assignment, students should have ample opportunity to develop a classroom community, with ongoing negotiations and debates to which they can speak, shared observations they might reference, and common terminology they all understand similarly.
- If you opt to have students write to a different audience (e.g., 5th-graders in the Flame Challenge, an editorial, a practitioner journal), then they should have ample opportunity to understand those communities.
- Assess ideas first: If you—as a fellow member of this course, with the knowledge you have—can clearly understand the idea(s) the student is writing about, the writing has done its first and most important job. Honor that in your feedback and in your grading.
- Assess consistency with classroom ideas: If an idea is consistent with the discussions you've had as a class, also honor that in your feedback and grading.

Part III

ADAPTATIONS

Bringing Writing Strategies into Different Settings

CHAPTER 9

Adaptations for Other Settings

> I have learned that I am terrible at grammar, but that I am capable of organizing my thoughts and ideas more clearly. Also that there is no such thing as having only one revised paper, but that I need to constantly revise.
> —Melanie

In this book, we advocate for writing in science classrooms that replicates the work that writing does in scientific communities. To do this kind of writing, students must be the authors of their own scientific ideas (they must have something to say), their classmates have to provide feedback and critique necessary for the development of those ideas (they must have someone to say it to), and this must be an iterative process that happens over extended periods of time.

Our class has certain structures—independent of our own instructional choices—that facilitate the work. Because our course is for students who are not science majors, and not preparing for an exam (i.e., the MCAT, Praxis, or GRE), the course has few traditional content requirements; defining focus as a "Seurat spot reunification point" is a perfectly acceptable outcome. This course also is designated a "writing proficiency" course by the university, so we are able to limit enrollment to 24 (while the physics and biology courses we teach are capped at 96). And, because we had the flexibility, we set it up as a lab course meeting for 4-1/2 hours a week, in a room with flexible seating arrangements.

We recognize that those affordances are rare. Many readers will need to meet a predefined set of content outcomes, often in very precisely defined ways that allow little room for student-generated representations and terminology. Your course may have requirements about scientific reading and writing outcomes, perhaps requiring that, by the end of the course, students will know how to prepare a journal article or how to use a database to select peer-reviewed scientific articles relevant to a topic. Or it may be a course with large enrollments, and tailoring exams to groups' ideas will not be practical, nor will whole-class discussions have the same ability to reach consensus. For those in K–12 settings, the list of standards is large, class sizes are increasing, and time is short. We hope that the ideas in this book provide leverage to advocate for flexible representations of content (we would argue that our students' definition of focus is as rigorous and accurate as any, even if it is noncanonical), a more nuanced interpretation of what "counts" as scientific writing, and smaller classes with more time for lab work in groups. Nonetheless, we know that an exact replication of our course is not likely. In this chapter, we offer suggestions for faculty who find themselves in other settings and hope to adapt some of the writing activities in this book. We focus primarily on how to allow for the following: students' iterative development of scientific ideas, the role of peer feedback in that development, and structuring writing activities to support that development.

LARGE-ENROLLMENT CLASSES

In 2009, one of the authors, Kim, designed a large, first-year writing course consisting of 90 students, nine student mentors, and the instructor. At her institution, California State University, Chico, they refer to this class as "the Jumbo" (Jaxon, 2014; Jaxon & Fosen, 2012). Three key structures support the overall design of the Jumbo: a small workshop that meets "outside but alongside" the large class (Grego & Thompson, 2008), writing mentors who participate in the course and lead the workshop component, and digital platforms that create a virtual "thirdspace" (Soja, 1996) to support student participation. All of these structures work together to support emerging scholars, mentors, and writers.

While we would not want to ignore the hard-fought battles for class-size reduction that writing instructors, K–12 teachers, and other educators have championed, the work in Kim's large writing class forces us to reconsider the arguments we hear when we talk about class size: Large classes certainly can make it challenging to give extensive feedback or even simply to know everyone's name. But class size is not the only variable that prevents students' ideas from being heard, engaged with, and refined; a classroom with 10 graduate students who never get a chance to speak because of a droning professor—who sees students as empty vessels to fill—can feel just as impersonal as a large lecture hall filled with people. However, these challenges are not impossible to overcome. The problem to address is how to create a space where a student feels like a valuable member, restructuring the class so a student is not anonymous and can receive feedback from peers and instructors. The structures of the large class, particularly the use of small teams led by mentors, make Kim's large class feel small. Students consistently respond in exit surveys that "we can't hide in this class and our ideas are taken seriously."

We would suggest modifying for large class size by imagining the structures, activities, and identities that are afforded in current classrooms and curricular designs—large or small—and considering how those structures, activities, and identities need to be re-imagined depending on class size. In other words, one of the problems when faculty try to scale a course is that they don't change the structures to account for the size: They simply do the same thing with more people. For example, as class size increases, are there other ways to share, challenge, and refine scientific ideas besides whole-class conversations? Kim uses a range of structures—blogging, live tweeting, silent discussions, and gallery walks, to name just a few—as a way to support participation beyond talking in class. She also places attention on developing students who are thoughtful peer responders of one another's work, thereby sharing the responsibility for feedback with the whole class. She knows she cannot give feedback to every piece of writing, and she trusts that students can give constructive feedback to one another and, more important, that the practice of giving feedback will help them in their growth as writers.

The large writing class allows for new ways to imagine community, participation, and the role of instructors. We see this redesign challenge as engaging and full of potential that leads to the creation of classrooms where students' ideas are valued. We can imagine a science course, modeled on this Jumbo course, that uses similar structures: small research teams within the large class, peer mentors, and/or digital spaces that allow teams to share work with the larger community.

In the large-course design, Kim uses a framework culled from game design, the concept of "epic" (McGonigal, 2011), as a mechanism to bring all of the structures of the large class—workshops, mentors, and digital platforms—together. She equates this to experiences in daily life: Sometimes we prefer to be at a rock concert with 10,000 people yelling the same lyrics at the top of their lungs, and other times we prefer a small club with only close friends listening to an acoustic set. One experience is not better or worse, but simply different. What McGonigal calls "epic scales" are those moments in which we recognize that the projects and actions we engage in and environments where they take place seem "bigger than ourselves" (p. 98). While many examples in McGonigal's work are drawn from online games like Portal and Halo, her argument extends to crowdsourcing and "real-life" applications like Foursquare, or distributed computing platforms that create protein-folding simulations to search for actual cures to Alzheimer's or Huntington's disease. Epic scales provide contexts for action as a form of service to these larger goals, encourage wholehearted participation, and—perhaps most relevant to our goals—provide mechanisms for the exchange of expertise. When systems are designed to help people share their interests and goals, she argues, people can be called upon and are motivated to do work they excel at. "And the chance to do something you're good at as part of a larger project helps students build real self-esteem among their peers," McGonigal says, "not empty self-esteem based on nothing other than wanting to feel good about yourself, but actual respect and high regard based on contributions you've made" (pp. 130–131).

Based on what we've learned from Kim's large, first-year writing course, we offer a few ideas for consideration in the design of a large, inquiry-based science class.

- ***Space matters:*** We scour our campus in search of classrooms that allow for student movement in large and small groups. We would argue that most large, lecture classrooms are set up for teaching, not learning. We were able to advocate for large rooms with rolling chairs. You may have similar spaces on your campus or can propose the design of alternative and configurable spaces.
- ***Group size:*** We find it useful for students to join permanent, small teams. We would also suggest asking students to create "norms," expectations they have about how they will work together as a group. These norms—guidelines—for their work together can be easily placed in a shared Google Doc or folder for reference when problems arise. Kim has found that the norms give students some control over how groups function, which in turn makes the groups higher functioning. Her students refer to the norms as "conditions in which someone can get kicked out of the group." Sometimes, students need to be able to hold one another accountable for their contributions (both too many contributions—a "bossy" group member—and too few).
- ***Mentor opportunities:*** We invite upper-division and graduate students to work with us in our large course. Many of these students are interested in careers in teaching, so the class provides a professional development opportunity for them. The mentors meet with the instructor for 2 hours outside of class to plan, read student writing, and respond to student drafts together.
- ***Digital tools:*** As a way to support large-class participation, we use a variety of digital platforms: Google Docs, Wordpress, Twitter, Facebook, to name a few. Students use these tools for group collaboration and as a way to share their ideas in low-stakes environments. In Kim's writing class, each small group designates one or two people for a "Blog Review Board," a small team of students with whom she meets. Instead of blogging, this team reads and responds to the blogs of the other students. Each week, this team nominates featured blogs to be showcased in the course. Participation in the review board allows for rich discussions about what makes "good writing," which this team then shares with their small group back in the large class. We can imagine setting up a similar structure where students are invited to send a team representative each week to confer with faculty about homework or other drafts. This team could bring back "featured homework," for example.

- ***Participation roles:*** We consider a variety of ways students can become productive members of our class. We should note that the roles students take up should emerge as organically as possible, meaning that they are offered and then taken up by students or not. Kim does not keep track of "how many times a student tweets" in a semester, for example; she offers Twitter as one place to share ideas, one way to contribute to the ideas of the course. As a class, we often return to roles and platforms to consider whether they are still working for us. We would not expect students to participate in every possible structure; the goal is for students to find a way to become valuable members of our class community. Some possible roles students can take on in a large class are:
  - Twitter "leads" in each of the small groups: One student in each team is responsible for tweeting out what is happening: interesting conversations, links people are finding, questions their conversations are raising, and so on. We use a class hashtag to follow our Twitter feed, and often we project the feed so we can all see the conversations happening around the large room. Kim often points to a tweet and then asks the team to expand on the idea for the whole class.
  - Visual and textual note-takers: A handful of students are responsible for keeping notes or creating visual representations of readings or ideas. They can keep notes and import images in a common Google Doc.
  - Group discussion leads for scientific ideas: Teams rotate the responsibility of developing key questions about the ideas in the course.
  - Master class: The reason for a master class in music is that not every musician who wants a lesson with a "master" can have one, and so we create a different structure to accomplish this. The master class easily translates into larger classes (for more on structuring a master class, see Chapter 2).

Ultimately, when thinking about large courses, we consider what elements of the class structures can be shared and distributed among students. Similar modifications are not foreign to large-enrollment science courses. Reform curricula in science over the past decade have rethought the structures in these courses to allow for more interactive engagement. SCALE-UP, in particular, offers suggestions for structures to create smaller cooperative learning groups (Beichner et al., 2007). The Learning Assistant model offers a model for developing a program that provides student mentors (learning assistants) in science classrooms (Otero, Pollock, & Finkelstein, 2010).

ADAPTATIONS FOR COURSES WITH PRESCRIBED CONTENT GOALS

Most science courses, of course, have well-defined content goals. Although the products of our students' inquiries are, in general, consistent with scientific ideas, they are nonetheless idiosyncratic (Atkins & Frank, 2015). For example, a typical introduction to lenses in a physics course will emphasize how to use the "lens maker's equation": using the focal length of a lens together with the object's location to calculate a position for a focused image. Students are given precise steps for drawing ray diagrams to find the size and location of an image, drawing three particular rays, and finding where they converge. Our students rarely develop such mathematical precision with their diagrams. They can describe why lenses must be more curved to focus on close objects, and flatter to focus on distant objects; they can offer evidence to suggest that objects are not in focus in our periphery; they have arguments for why a blue iris should have a more "washed out" image on the retina than a dark iris; and one group constructed an explanation for why images look blurry underwater, essentially arguing that the index of refraction for water was too similar to that of a lens for the eye to focus.

When comparing those ideas with the kinds of ideas that are developed in an introductory physics course, we find them to be as rich and complex as any more standard description of lenses. When possible, we urge faculty to consider ways to "count" these outcomes as legitimate content outcomes and expand what we understand to be content.

However, we know that most faculty—particularly those in K–12 settings—have a more narrowly defined description of what students must learn, for example, about lenses. In each chapter, then, there are pieces that can be used in a more traditional classroom. For example, the instructor can:

- Work through a homework problem as a master class (Chapter 2) and draw out students' ideas about not only a particular answer, but a deeper understanding of the science behind that answer. This models for them how to work with one another on problem sets and the ways that they can engage with homework.
- Have groups work together on a particularly challenging representation, sharing their diagrams on whiteboards with one another and iteratively improving them (Chapter 3); and, when possible, consider affordances of noncanonical representations with students, while also ensuring that they are familiar with standard representations.
- Offer students an opportunity to engage with a problem before reading the textbook (Chapter 5) and then use a Google Doc (or similar environment) to allow students to read "together" and use their comments to motivate instruction.

- Develop homework assignments that draw on student ideas (Chapter 6), positioning student ideas alongside the textbook's descriptions.
- Make peer feedback an ongoing, embedded part of students' work, not only in formal classroom structures (whiteboards, diagrams, and reading together), but also in homework, by facilitating groups to work together on homework (Chapters 2–7); and, in providing faculty feedback on homework, offer more than a simple right/wrong judgment of their work.
- Have students construct brief sentence summaries to refine in class, similar to the way that definitions are refined in Chapter 7.

A FEW EXAMPLES FROM OUR OWN COURSES

In our undergraduate physics labs, the students answer a set of questions at the conclusion of each lab. For each class, Leslie creates a shared Google Doc where lab groups each contribute their responses, react to other groups' responses, and through a whole-class conversation at the end of the class create a final draft of their conclusions. These documents are all available through the course's learning management system (Blackboard).

In guided-inquiry courses, Physics and Everyday Thinking (Goldberg, Robinson, & Otero, 2008) and Life Science and Everyday Thinking (Donovan et al., 2013), students construct ideas throughout a unit before reading a "Scientists' Ideas" document that asks them to compare their own ideas with canonical scientific ideas and to summarize the experimental evidence that supports those ideas. We import that document into a Google Doc and use it as we do the readings described in Chapter 5.

When students were learning about force, energy, and motion in a physics course, Leslie asked them to write a sentence (using force, motion, and energy) that describes what happens when you throw a ball. Students' sentences were relatively straightforward. The one we edited was: "As I throw a ball, before I release, I am applying energy by giving a force and velocity to the ball." To those familiar with physics, there are some terms that give us pause: Energy usually is not described as "applied" to an object. Instead, it is transferred to or from an object. Meanwhile, the ontology of force is such that saying it is "applied" makes more sense—or at least is less ambiguous—than saying it is "given." *Give* means so many kinds of things: cause to have ("she gives me a headache"), transfer to ("I give her my money"), and interact via ("I give her a handshake"), to name a few. We discussed that ambiguity and then rewrote the sentence without *give*, paying careful attention to the nature of the interactions, and reflecting on prior work and definitions that had come before. Ultimately, students rewrote the sentence as: "I apply a force to the ball by pushing it, which transfers some energy, and that (a transfer of energy) causes it to have motion."

We have argued in prior chapters that students are willing to do this challenging work that requires personal intellectual investment because they know it will be noticed, valued, and used in class. It is more difficult, then, for students to invest themselves in developing their own ideas when they know that they will be held accountable to very particular scientific ideas and representations regardless of their own commitment to those ideas. Instructors who incorporate these activities also should ensure that the products are valued as the class moves forward.

ADAPTATIONS FOR K–12

All of the authors have experience working in K–12 settings: Irene recently left higher education and is an elementary school principal, Leslie has led science inquiry units for an elementary school, and Kim has worked with student writing in a variety of K–12 contexts through her involvement with both the National Writing Project and the Northern California Writing Project. As an example of the work we've done with younger students, Leslie, in teaching 2nd-graders about decomposition, began by having students fill Tupperware containers with food and leaving them (sealed) in the classroom for a few weeks. These young writers drew pictures in their lab books of the changes that happened and, using some sentence stems to help them, wrote down their questions. Students wondered why the food "shrunk," where the "white and fuzzy stuff" (molds and bacteria) came from, and why there were bubbles. Child-friendly books (e.g., Cobb, 1981) were introduced as we read together and related the book's descriptions of decomposition to our own observations and questions. "Maybe the bacteria are blowing bubbles," one student noted. Motivated by the book, students went on a "microbe hunt," choosing something they thought might have microbes on it and cultivating those in petri dishes, and, finally, discussing why some things had more microbes than others.

There are many books that offer examples of early elementary students "composing science." For instance, Buhrow and Garcia (2006) and Whitin and Whitin (1997) take readers into their classrooms, with a rich blend of science and literacy practices. We often assign readings from Gallas (1995) to future elementary teachers as they are taking our inquiry course. And the online resource of Austin's Butterfly (EL Education, n.d.) offers an example of how to engage students (here, 1st-graders) in iterative feedback and drafts in scientific representation, not at all unlike our work with diagrams.

In elementary school, Irene has piloted science and literacy units on a variety of student-selected topics, with 2nd- and 3rd-graders using science notebooks, whiteboards, short writing assignments, and readings drawn from popular science news articles. For example, one unit began with students brainstorming science topics that they were curious about. Students rallied around the idea of exploring Mentos and Diet Coke reactions. Perhaps it is related to the "strength" of the mints? But Altoids made fewer bubbles than mild Mentos. Could it be related to the addition of sugar (Mentos) to a sugar-free drink? This led to students dropping in sugar cubes and reading nutrition labels to compare sugar content across sodas and mints. Ultimately, however, sugar was not the reason. They wondered, then, whether it was a chemical reaction like baking soda and vinegar. After a week of exploring the differences between chemical and physical changes, students concluded that the explosion was a physical change. After they had developed their own ideas through the iterative processes of writing, sharing, discussing, and refining, Irene introduced the work of scientists on this issue into the classroom conversation (adapted from the strategies described in Chapter 5) by asking them to read an article drawn from *New Scientist,* a popular science magazine (Muir, 2008). Afterward, these 8- and 9-year-olds begged for a copy of the original research paper (Coffey, 2008), where they homed in on the data table and carefully compared their own class's data with those of the scientists.

As students enter middle and high school, they have more content knowledge to draw from in developing scientific explanations and have a richer knowledge of genres. Classes are expected to include a strong lab component, with many high school science classrooms set up to accommodate inquiry. The changes heralded by the NGSS and Common Core allow for more attention to inquiry practices in science and position writing instruction as a central part of disciplinary teaching outside of English class. And faculty often have a more extensive background in science and feel more confident in managing the range of ideas that can emerge through inquiry. At the same time, it is in middle and high school that students' writing often becomes increasingly constrained, with emphasis on forms such as the five-paragraph essay, structured lab reports, and science fair presentations. In addition, the content requirements become increasingly complex, and stakes—exams, graduation, and college on the horizon—may seem higher. In part because of these demands, students report enjoying science and writing less than they did in the earlier grades. Teachers, too, have constraints that may limit the amount of inquiry and writing they find manageable: They usually teach many more students, and for shorter class periods, with a longer list of content standards. Their strong scientific backgrounds rarely include the kinds of literacy methods background that elementary teachers have. That is, while students and teachers in secondary schools have more practices and content to draw from, these constraints can make it difficult to address literacy instruction in science courses in secondary schools.

Having taught in secondary schools (Irene in a middle school and Leslie in high schools), we find labs to be a rich opportunity to rethink traditional approaches and incorporate more writing instruction. Many faculty, of course, are already doing this with support from new standards and emphasis on inquiry instruction. Rather than prescribed, step-by-step labs in which students are asked to demonstrate a known law, we now engage students in labs that ask them to use observations to develop and refine models. Explicit attention to the role of notebooks, whiteboards, and diagrams in this process is given in service of the scientific ideas, while also developing writing practices. Lab write-ups that provide opportunities for feedback and iteration, and move away from prescribed structures toward ones that students find useful, can facilitate the production of meaningful texts and an understanding of the kinds of choices that authors may make. And engaging in reading scientific texts together, after students have developed their own ideas, interests, and questions, can replicate the work that we describe in Chapter 5, without the requirement that all students have access to the Internet at home.

Take-Home Messages

- Writing in science is far more than the final published article; as learners, students will gain more from the more informal, social, and iterative writing practices that scientists use when they are developing ideas. We find an emphasis on using writing to develop, share, and vet ideas, regardless of the context, to be more important than high-stakes writing assignments that summarize final ideas.
- Feedback, feedback, feedback. Ideas are improved through ongoing, iterative feedback by peers and faculty. Writing is improved as ideas are improved, and it is through a range of informal and formal feedback structures that students garner feedback.
- Structures for participation—the whole-class conversation, consensus definitions, methods of peer feedback, generating lab notebook rubrics—must change as classroom contexts change. There are affordances that can be engaged in each context (e.g., increasing emphasis on peer feedback, the "epic" scale of participation).

CHAPTER 10

Addressing Content Standards in the Era of NGSS and Common Core

> It gives me the chance to really get my mind out in the open and discover how far I can dig into millions of questions to have my own answers instead of somebody else to lead me to it. Also, it makes me feel good about myself that I know I did something on my own. —Susanna, 7th-grader

In the Introduction, we noted that the approaches to writing we address in this book are consistent with standards in K–12 education: the Next Generation Science Standards (NGSS) (NGSS Lead States, 2013) and the Common Core State Standards for English Language Arts (ELA) & Literacy in History/Social Studies, Science, and Technical Subjects (Common Core State Standards Initiative, 2010).

Central to the NGSS is the idea that we learn scientific ideas by engaging in scientific practices. No longer is there an "inquiry unit" that stands apart from other, more content-focused units in science class: Inquiry—as described by the eight Scientific Practices—is *how* students develop strong conceptual understanding of scientific content (Disciplinary Core Ideas). And new to the Common Core is a greater emphasis on nonfiction reading and writing than historically has been part of language arts instruction. The Common Core ELA Standards emphasize the roles of reading, writing, and speaking-and-listening to understand, critique, and communicate academic content outside of English class. Like the eight Scientific Practices that are engaged in the development of scientific Core Ideas in the NGSS, the practices of reading, writing, and speaking-and-listening are employed in developing the Anchor Standards for college and career readiness in the Common Core.

Both the NGSS and Common Core describe the kinds of tasks that students should be able to do, as opposed to prescribing particular methods or lessons that develop these results. In other words, the standards highlight what students should be doing, but they do not provide specific curriculum or a how-to for getting to those desired outcomes. This allows teachers to cull from their own expertise and strengths, and, together with a knowledge of their students, to design instruction to meet the standards. In many ways, this book complements the efforts of the new standards: We focus on methods that we use in our courses—providing some suggestions for structures and activities—but without close attention to particular content (the disciplinary core ideas of the NGSS) that students develop with these methods.

In this chapter, we provide connections between the desired outcomes—as represented in the NGSS and Common Core—and the processes we have suggested throughout the book. We explore the standards in relation to each chapter's focus in order to demonstrate that the work of this book is also the work of the NGSS and Common Core. Our hope is that, for example, the science teacher who has students comparing a range of student-constructed definitions on whiteboards will be able to support this instructional choice by references not only to scientific practice, as described in prior chapters, but to the Scientific Practices in the NGSS. And the English teacher who brings lenses into his classroom as students develop representations for magnification can relate the iterative development of representations not only to scientific work, but to the anchor standards described in the Common Core.

Below we briefly outline the Scientific Practices of the NGSS, followed by a short overview of the Anchor Standards in the Common Core. We then turn to each chapter in the book, summarizing its main themes and highlighting one practice or standard that is particularly relevant to that chapter. The range of NGSS engaged in, say, a lab notebook is vast—the writing will be connected to disciplinary core ideas, cross-cutting themes, and a range of scientific practices. The strands of the Common Core ELA Standards addressed in a lab notebook are similarly large: Notebooks are places of writing, are revisited as students communicate ideas, are sites of language development, and, at times, are shared and read by others. Each of the examples below, then, provides one particularly salient standard to consider; this list is by no means exhaustive.

NGSS: SCIENTIFIC AND ENGINEERING PRACTICES

The Next Generation Science Standards have three "dimensions": Practices, Disciplinary Core Ideas, and Cross-cutting Concepts. The disciplinary core ideas are the topics that commonly have been thought of as the "content" of science courses: information about the structure of the atom, the water cycle, or cellular respiration, to name a few. The cross-cutting concepts (e.g., patterns and cause and effect) are "common and familiar touchstones across the disciplines and grades"—recurring themes that transcend disciplinary boundaries and create a coherent understanding of scientific content. Both of those dimensions will be specific to the content that you teach. The practices, however, underlie the development of all scientific content. These are:

Practice 1: Asking questions (for science) and defining problems (for engineering)
Practice 2: Developing and using models
Practice 3: Planning and carrying out investigations
Practice 4: Analyzing and interpreting data
Practice 5: Using mathematics and computational thinking
Practice 6: Constructing explanations (for science) and designing solutions (for engineering)
Practice 7: Engaging in argument from evidence
Practice 8: Obtaining, evaluating, and communicating information

The NGSS practices are not intended to be thought of as separate skills that students move through in a linear fashion. The practices overlap, weave, and build on one another as students come to understand the work of science (and engineering) and its use in solving problems. Even the naming of this list as "practices" (as opposed to skills) honors the ways in which learning to do anything requires an understanding of how the doing is embedded in a community and is not a rote activity.

COMMON CORE: ANCHOR STANDARDS

The English Language Arts Standards are divided into four separate areas: Reading, Writing, Speaking-and-Listening, and Language. Consistent with the examples in our book, in which students may read and comment on texts before engaging in a discussion about a definition (thereby engaging in all four areas: reading, writing, speaking-and-listening, and language), the authors of the ELA Standards note that the separate areas are "for conceptual clarity," and "the processes of communication are closely connected."

Within each strand are anchor standards, the long-term goals of ELA instruction. While a 2nd-grade teacher should not expect her students to meet Language Anchor Standard 1 ("Demonstrate command of the conventions of standard English grammar and usage when writing or speaking"), these targets are in mind when considering the standards that 2nd-graders should meet. We will address the anchor standards below, but note that different grades will be advancing toward those standards in different, developmentally appropriate ways.

Chapter 1: Student Notebooks

All formal scientific writing begins as informal scribbles on a page. We create a space in our courses—the lab notebook—that respects this initial step in the process. In developing criteria for the lab notebook, we begin with reading scientists' notebooks, dismantling some myths about the scientific process: that it proceeds in an orderly and uniform manner, from hypothesis to conclusion. By having students develop the criteria for their notebooks—using scientists' notebooks as a template—we not only give students ownership in the assessment process but foreground the central role that students' assessment will play in our class.

Standards

As noted in the foundational document for the NGSS, *A Framework for K–12 Science Education* (National Research Council, 2012, p. 44), "a focus on the practices (in the plural) avoids the mistaken impression that there is one distinctive approach common to all science—a single 'scientific method.'" In this way, our work with notebooks highlights the idiosyncratic, creative, and nonlinear processes of scientific inquiry.

More specifically, within the Scientific Practices, the notebook is a medium where many of the practices take place. Consider the first four practices:

Practice 1: Asking questions and defining problems
Practice 2: Developing and using models
Practice 3: Planning and carrying out investigations
Practice 4: Analyzing and interpreting data

Note that the practices use a gerund construction: asking, defining, developing, using, planning, carrying out, analyzing, and interpreting. These are ongoing, iterative activities—they happen while students are discussing ideas, reading, experimenting, and reflecting on those experiments. The lab notebook is a place where many of those practices become text on the page for the first time.

Chapter 2: Whiteboards

In contrast to the relatively personal lab notebooks, which hold individual students' ideas and reflections, the whiteboards serve as a more public place where competing ideas are placed side by side and subjected to more formal scrutiny. It is writing in which students are coordinating their group's data, their individual ideas, and information from other groups and texts, and succinctly summarizing those to share with peers. They interpret others' whiteboards—learning how to "read" these, develop conventions, critique others' work, and using others' critiques to further their own work. Unlike their notebooks and homework assignments, these ideas necessarily are negotiated by the lab group prior to being shared with the whole class. It is a site of ongoing, iterative but informal feedback on ideas, in the way that scientific lab groups and colleagues sketch and refine ideas together.

Standards

Although many scientific practices are engaged when developing and discussing whiteboards, here we turn to the Common Core ELA Anchor Standards—in particular, the strand related to speaking-and-listening. This strand is summarized as follows:

> Students must have ample opportunities to take part in a variety of rich, structured conversations—as part of a whole class, in small groups, and with a partner. Being productive members of these conversations requires that students contribute accurate, relevant information; respond to and develop what others have said; make comparisons and contrasts; and analyze and synthesize a multitude of ideas in various domains.

The two anchor standards that are particularly well-aligned with our use of whiteboards include:

1. CCSS.ELA-LITERACY.CCRA.SL.4: "Present information, findings, and supporting evidence such that listeners can follow the line of reasoning and the organization, development, and style are appropriate to task, purpose, and audience."
2. CCSS.ELA-LITERACY.CCRA.SL.5: "Make strategic use of . . . visual displays of data to express information and enhance understanding of presentations."

As noted above, these anchor standards are end goals for instruction—what a graduating senior should be able to do. The ways in which we use whiteboards in class to share our ideas give students ongoing practice and feedback, as they become more proficient in presenting their findings and using visual displays.

Chapter 3: Diagrams

"Writing" has come to mean all of the ways in which we use inscriptions: We compose not only texts, but a range of representations to convey scientific ideas. As with all writing, these representations begin informally and progress through an iterative process with feedback from peers and with attention to our expanding base of experimental evidence. While science relies on shared, uniform representations for clarity (e.g., the Lewis dot structure and phase diagrams in chemistry, free-body diagrams and light ray diagrams in physics, conventions around graphing, tables, etc.), and instruction on those representations is important, students also need practice in constructing (and refining) novel representations to convey their own ideas and making their own representational choices based on the information they hope to highlight.

Standards

Both the Common Core and NGSS highlight the importance of constructing and interpreting diagrams, tables, and graphs. We turn back to the NGSS practices in describing the ways in which our work with diagrams is connected to K–12 standards. In particular, Practice 8: Obtaining, evaluating, and communicating information, is addressed with the iterative development of diagrams. By grade 12, according to the NGSS, students should be able to "use words, tables, diagrams and graphs . . . to communicate their understanding." Scientific texts are incredibly multimodal (Kress & van Leeuwen, 2001), and any attention to scientific writing necessarily will need to support the construction of a range of representations.

Chapter 4: Peer Review

In Chapter 4, we note that writing well is a process of iteration. In science, as in all academic fields, we write, rewrite, and rewrite as we move from initial ideas in a notebook, to shared understandings on a whiteboard, to drafts of our conclusions, and to a manuscript that, with more formal feedback, evolves into a published article. The changes and improvements that happen with each revision are based on rigorous self- and peer feedback. Participating in giving and receiving feedback is, then, fundamental to participation in the construction of scientific ideas. That is, to review well, you must assess ideas well—understanding the claims, weighing them in light of competing ideas and evidence, and conveying that assessment to the author. And so, in assessing others' scientific ideas, students become increasingly skilled at many scientific practices.

We do not reserve peer review for final papers or drafts of scientific essays. Instead, we hope that our structures support a stance toward idea development, a shared interest in developing students' own and their peers' ideas. Peer review is one way communities are formed: When a peer takes an idea seriously, seriously enough to offer kind but critical feedback, we feel valued.

Standards

Both the Common Core and NGSS ask students to work with multiple claims through comparing and contrasting evidence and findings. The Common Core offers the following standard that speaks directly to the work of peer review and idea development: "Assess the extent to which the reasoning and evidence in a text support the author's claim or a recommendation for solving a scientific or technical problem" (CCSS.ELA-LITERACY.RST.9-10.8).

The language of the NGSS is almost identical to the Common Core standards with regard to developing the practices of synthesizing and analyzing competing ideas. The NGSS, for example, ask students to "compare and contrast data collected by different groups in order to discuss similarities and differences in their findings" (Practice 4). Silent science and gallery walk activities, described in Chapter 4, support this practice.

Chapter 5: Reading Together

Chapter 5 is perhaps the most closely aligned with ideas in the Common Core and NGSS. When most faculty consider what it means to teach students about reading in technical fields, what comes to mind is instruction that addresses understanding and building on ideas from texts produced outside of the classroom.

In this book, we argue that scientists read "together." As reviewers, editors, and participants in journal clubs and lab groups, scientists discuss and make sense of scientific writings with one another; similarly, readings for class are part of the social meaning-making that students do as they consider scientists' ideas and implications of those ideas. Moreover, reading is goal-directed and related to the ongoing questions that scientists grapple with.

Standards

The following standard is particularly relevant to our course and our use of text: "Evaluate the hypotheses, data, analysis, and conclusions in a science or technical text, verifying the data when possible and corroborating or challenging conclusions with other sources of information" (CCSS.ELA-LITERACY.RST.11-12.8).

This evaluation is not in the style of a "book report" or other synopsis provided to the instructor by the students, but instead is a conversation that takes place between students in the context of their own ongoing scientific discovery. They read to determine whether and how the text speaks to their own work. When we read texts, students generally have already developed their own hypotheses and data, begun some analysis, and generated nascent conclusions; these provide the other sources and conclusions against which they weigh the text. We "verify" the data in the way scientists often do: not by replicating experiments, but by a critical reading of the methods that were employed, and the plausibility of the outcomes. We demonstrate this in more detail in Chapter 5.

As with all NGSS Practices, those related to reading in the sciences are intended to be employed "three-dimensionally," with attention not only to the practice, but to the disciplinary core idea(s) and cross-cutting concept(s). That is, by engaging in scientific practices (like critically reading scientific texts), students develop content knowledge (disciplinary core ideas) in areas that are broadly relevant across the sciences (cross-cutting concepts).

The NGSS practices related to reading are similar to those in the Common Core ELA Standards, but have one notable distinction: The NGSS ask students not only to make sense of complex, technical writing, but to use that to "address a scientific question or solve a problem." That is, in the context of a science class, it is not enough to analyze scientific writing, but students should use it to further their own work, as part of a broader set of practices that support and sustain scientific inquiry.

Chapter 6: Homework

The theme of this book is that the writing students do in science courses should more directly replicate the work that writing does in the sciences: often informal, always iterative, inherently social. Homework, on its surface, has little connection to scientific practice. However, that iterative, social progression of ideas requires that students have time to make sense of the ideas and evidence that have emerged. And so we use homework in a way that weaves together the developing narrative of scientific ideas in our class: attending to the competing theories, models, and evidence that have been introduced during class, and ones we hope to refine in the days ahead. It is an activity that supports newcomers as they learn to participate and prepare to be members of a scientific community.

Participation in homework, then, although completed on one's own, is a requirement for full participation in the class. It is one more way in which the writing—done individually—nonetheless is embedded in a social context.

Standards

The Anchor Standards are intended to paint a broad picture, while the grade-level standards in reading, writing, and speaking-and-listening offer specifics in each of the main ELA areas. A few key writing standards that are addressed in homework assignments include: "writing arguments to support claims" (CCSS.ELA-LITERACY.W.11-12.1), "writing informative/explanatory texts to examine and convey complex ideas, concepts, and information clearly and accurately" (CCSS.ELA-LITERACY.W.11-12.2), "strengthen[ing] writing as needed by planning, revising, editing, rewriting, or trying a new approach, focusing on addressing what is most significant for a specific purpose and audience" (CCSS.ELA-LITERACY.W.11-12.5), and "conducting short as well as more sustained research projects to answer a question" (CCSS.ELA-LITERACY.W.11-12.7). One additional standard stands out as particularly relevant to the role that homework plays in our course: "Prepare for and participate effectively in a range of conversations and collaborations with diverse partners, building on others' ideas and expressing their own clearly and persuasively" (CCSS.ELA-LITERACY. CCRA.SL.1). That is, it is through homework that students prepare for the range of conversations that will take place in class.

Chapter 7: Definitions

Chapter 7 focuses on constructing brief, polished "consensus" statements: statements that summarize hard-won ideas succinctly and clearly, and often—in our push for clarity—launch new questions. Understandably, the Common Core and NGSS do not take a position on the length of students' writing. However, these brief statements are important in replicating the kinds of writing that those documents call for. While these statements are not necessarily persuasive or argumentative pieces in themselves, a clearly stated definition is critical for many scientific arguments. And although these statements may not explicitly refer to multiple texts or data, they represent a synthesis, drawing for their development on a range of ideas and experiments: diagrams, data, and models. So these brief statements—which often make no explicit reference to empirical work, and are not complete arguments in themselves—are pedagogically useful in that they help support longer, more complex writing and can even help structure experiments and data analysis.

Standards

The process of constructing a definition is most closely related to the practice of obtaining, evaluating, and communicating information: "Scientists and engineers must be able to communicate clearly and persuasively the ideas and methods they generate. Critiquing and communicating ideas individually and in groups is a critical professional activity." Definitions play into this standard in several ways: (1) they must themselves be clear; (2) a clear definition is an important piece of persuasive arguments; and (3) our students critique and communicate in groups in a way that mirrors professional activity.

In addition, the Common Core standards note that students should "write routinely over extended time frames (time for research, reflection, and revision) and shorter time frames (a single sitting or a day or two) for a range of tasks, purposes, and audiences" (CCSS.ELA-LITERACY. CCRA.W.10). These brief, summative writings, produced in a group, are illustrative of the "shorter time frames" with a specific purpose in mind.

Chapter 8: Final Papers

In Chapter 8 we argue that to write well an author must have something to say and someone to say it to. And we argue that the work described in earlier chapters is in service of those goals: It is through the iterative development of their own scientific ideas that students develop "something to say." And it is through the ongoing interactions with their peers and instructors, who challenge, contradict, support, and delight in those ideas, that they develop "someone to say it to." In this way, disciplinary core ideas—the second "dimension" of the NGSS—are developed and also personalized.

Standards

These final papers support a thread of standards in the Common Core strand related to writing. In particular, this writing, which summarizes findings from many students in the class, together with any texts we've used, over weeks of instruction, supports the requirement that students can "write informative/explanatory texts to examine and convey complex ideas and information clearly and accurately through the effective selection, organization, and analysis of content" (CCSS.ELA-LITERACY.CCRA.W.2). In addition, our final papers address the strand related to building and presenting knowledge through writing: "Conduct short as well as more sustained research projects based on focused questions, demonstrating understanding of the subject under investigation" (CCSS.ELA-LITERACY.CCRA.W.7).

FINAL CONSIDERATIONS

Although certainly not necessary (we can imagine addressing these standards with more traditional readings), most of the "texts" in our class are the presentations, assignments, and ideas from our students; the "authors" that are being analyzed are, primarily, their peers; the "primary source" texts are their own data and observations. While we certainly introduce texts from scientists (see Chapter 5), students are wary of critiquing Newton. They assume, quite reasonably, that Newton's ideas are correct, and it would be gutsy (if not arrogant) for a student to challenge Newton's data and assumptions.

That said, we believe that students learn to write well, in part, by reading well-written pieces. We learn to construct beautiful diagrams, in part, by examining high-quality diagrams from our field. We would not argue that analyzing other students' work is sufficient for developing as a writer. However, it is essential that students learn to make their own representational choices, analyze and construct their own explanations of data, critique preliminary ideas, and synthesize findings. This is facilitated in our course by the emphasis on students' own ideas; scientific texts come later. Even in more traditional lab courses, making space for these kinds of analyses is important.

As a final point, data function much like a "primary text" for this course—the observations that students make and bring to bear in constructing interpretations of the world and defending their claims. It is the same role that primary sources play in history and that a novel plays in literary analysis. So when the Common Core ELA Standards emphasize the role of texts in constructing arguments, syntheses, and analyses, we construe "text" broadly to include things like:

- Observations
- Inscriptions in students' lab notebooks and those of their peers
- Scientific readings: things students find and articles the instructors introduce
- Diagrams: both those that students construct and those that appear in readings

Similarly, the NGSS call on teachers to weave together disciplinary core ideas, cross-cutting concepts, and scientific practices. While the particular disciplinary ideas will vary from discipline to discipline and course to course, and the broader cross-cutting concepts (e.g., energy, patterns, or stability and change) will depend on the particular topic, the eight practices are relevant to all disciplines and topics.

References

Adler-Kassner, L., & Wardle, E. (Eds.). (2015). *Naming what we know: Threshold concepts of writing studies*. Logan, UT: Utah State University Press.

Arsenault, D. J., Smith, L. D., & Beauchamp, E. A. (2006). Visual inscriptions in the scientific hierarchy. *Science Communication, 27*, 376–428.

Atkins, L. J. (2012). Peer-assessment with online tools to improve student modeling. *The Physics Teacher, 50*, 489–493.

Atkins, L. J., & Frank, B. W. (2015). Examining the products of responsive inquiry. In A. Robertson, D. Hammer, & R. E. Scherr (Eds.), *Responsive teaching in science* (pp. 56–84). New York, NY: Routledge.

Atkins, L. J., & Salter, I. (2011, August). *"What's a fourth? What's a fifth? What's the point?": Constructing definitions in scientific inquiry*. Presentation at the annual conference of EARLI, Exeter, UK.

Ball, D. L. (1993). With an eye on the mathematical horizon: Dilemmas of teaching elementary school mathematics. *The Elementary School Journal, 93*(4), 373–397.

Bardovi-Harlig, K. (2000). *Tense and aspect in second language acquisition: Form, meaning, and use*. Oxford, UK: Blackwell.

Bartholomae, D. (1980). The study of error. *College Composition and Communication, 31*(3), 253–269.

Bartholomae, D. (1985). Inventing the university. In M. Rose (Ed.), *When a writer can't write* (pp. 134–165). New York, NY: Guilford Press.

Baxter, G. P., Bass, K. M., & Glaser, R. (2001). Notebook writing in three fifth-grade science classrooms. *The Elementary School Journal, 102*(2), 123–140.

Bazerman, C. (1988). *Shaping written knowledge*. Madison: University of Wisconsin Press.

Beichner, R., Saul, J., Abbott, D., Morse, J., Deardorff, D., Allain, R. . . . Risley, J. S. (2007). Student-Centered Activities for Large Enrollment Undergraduate Programs (SCALE-UP) project. In E. Redish & P. Cooney (Eds), *Research-based reform of university physics* (pp. 1–42). College Park, MD: American Association of Physics Teachers.

Brandt, D. (1998). Sponsors of literacy. *College Composition and Communication, 49*(2), 165–185.

Buhrow, B., & Garcia, A. U. (2006). *Ladybugs, tornadoes, and swirling galaxies: English language learners discover their world through inquiry*. Portland, ME: Stenhouse.

Cobb, V. (1981). *Lots of rot*. Philadelphia, PA: Lippincott Williams & Wilkins.

Coffey, T. S. (2008). Diet Coke and Mentos: What is really behind this physical reaction? *American Journal of Physics, 76*(6), 551–557.

Common Core State Standards Initiative. (2010). Common Core State Standards for English language arts and literacy in history/social studies, science, and technical subjects. Retrieved from www.corestandards.org

Cunningham, A. (1988). Getting the game right: Some plain words on the identity and invention of science. *Studies in History and Philosophy of Science, 19*, 365–389.

Davidson, C. (2010). 21st century literacies: Syllabus, assignments, calendar. Humanities, Arts, Science, and Technology Alliance and Collaboratory (HASTAC). Retrieved from hastac.org/blogs/cathy-davidson/2010/12/31/21st-century-literacies-syllabus-assignments-calendar

De Villiers, J. G., & De Villiers, P. A. (1973). A cross-sectional study of the acquisition of grammatical morphemes in child speech. *Journal of Psycholinguistic Research*, 2(3), 267–278.

Donovan, D. A., Atkins, L. J., Salter, I. Y., Gallagher, D. J., Kratz, R. F., Rousseau, J. V., & Nelson, G. D. (2013). Advantages and challenges of using physics curricula as a model for reforming an undergraduate biology course. *CBE—Life Sciences Education, 12*, 215–229.

Dunbar, K. (1995). How do scientists really reason: Scientific reasoning in real-world laboratories. In R. J. Sternberg & J. E. Davidson (Eds.), *The nature of insight* (pp. 365–396). Cambridge, MA: MIT Press.

Dunbar, K. (1999). The scientist in vivo: How scientists think and reason in the laboratory. In L. Magnani, N. Nersessian, & P. Thagard (Eds.), *Model-based reasoning in scientific discovery* (pp. 89–98). New York, NY: Plenum.

Ede, L., & Lunsford, A. (1984). Audience addressed/audience invoked: The role of audience in composition theory and pedagogy. *College Composition and Communication, 35*(2), 155–171.

Edgerton, S. (1985). The Renaissance development of the scientific illustration. In J. Shirley & D. Hoeniger (Eds.), *Science and the Arts in the Renaissance* (pp. 168–197). Washington, DC: Folger Shakespeare Library.

EL Education. (n.d.). Retrieved from modelsofexcellence.eleducation.org/projects/austins-butterfly-drafts

Elby, A. (2009). Defining personal epistemology: A response to Hofer & Pintrich (1997) and Sandoval (2005). *Journal of the Learning Sciences, 18*(1), 138–149.

Feuer, M. J., Towne, L., & Shavelson, R. J. (2002). Scientific culture and educational research. *Educational Researcher, 31*(8), 4–14.

Francek, M. (2006). Promoting discussion in the science classroom using gallery walks. *Journal of College Science Teaching, 36*(1), 27–31.

Gallas, K. (1995). *Talking their way into science: Hearing children's questions and theories, responding with curricula*. New York, NY: Teachers College Press.

Garcia-Mila, M., & Andersen, C. (2007). Developmental change in notetaking during scientific inquiry. *International Journal of Science Education, 29*(8), 1035–1058.

Gee, J. (1996). *Social linguistics and literacies: Ideology in discourses*. Philadelphia, PA: Routledge Falmer.

Giles, J. (2012). The digital lab: Lab management software and electronic notebooks are here—and this time it's more than just talk. *Nature, 481*, 430–431.

Goldberg, F. M., Robinson, S., & Otero, V. (2008). *Physics & everyday thinking*. Armonk, NY: It's About Time, Herff Jones Educational Division.

Grego, R., & Thompson, N. (2008). *Teaching/writing in third spaces: The studio approach.* Carbondale: Southern Illinois University Press.

Halmos, P. R. (1973). How to write mathematics. In N. E. Steenrod, P. R. Halmos, M. M. Schiffer, & J. R Dieudonné (Eds.), *How to write mathematics* (pp. 19–48). Providence, RI: American Mathematical Society.

Harnad, S. (1990). Scholarly skywriting and the prepublication continuum of scientific inquiry. *Psychological Science, 1*, 342–343.

Harris, M. (1995). Why writers need writing tutors. *College English, 57*(1), 27–42.

Jaxon, K. (2006). One approach to guiding peer response. *Program handbook for academic writing.* Chico, CA: California State University, Chico, English Department. Retrieved from http://www.nwp.org/cs/public/print/resource/2850

Jaxon, K. (2014). Jumbo course design. Retrieved from kimjaxon.com/?page_id=146

Jaxon, K., & Fosen, C. (2012). Not your grandmother's composition class. *NWP radio.* Retrieved from http://www.nwp.org/cs/public/print/resource/3977

Keys, C. W., Hand, B., Prain, V., & Collins, S. (1999). Using the science writing heuristic as a tool for learning from laboratory investigations in secondary science. *Journal of Research in Science Teaching, 36*(10), 1065–1084.

Kirschner, P. A., Sweller, J., & Clark, R. E. (2006). Why minimal guidance during instruction does not work: An analysis of the failure of constructivist, discovery, problem-based, experiential, and inquiry-based teaching. *Educational Psychologist, 41*(2), 75–86.

Klentschy, M. (2005). Science notebook essentials. *Science and Children, 43*(3), 24–27.

Kozma, R. (2003). The material features of multiple representations and their cognitive and social affordances for science understanding. *Learning and Instruction, 13*, 205–226.

Kress, G. (2009). *Multimodality: A social semiotic approach to contemporary communication.* New York, NY: Routledge.

Kress, G. R., & van Leeuwen, T. (2001). *Multimodal discourse: The modes and media of contemporary communication.* London, UK: Hodder Arnold.

Latour, B. (1990). Drawing things together. In M. Lynch & S. Woolgar (Eds.), *Representation in Scientific Practice* (pp. 19–68). Cambridge, MA: MIT Press.

Latour, B., & Woolgar, S. (1979). *Laboratory life: The construction of scientific fact.* Beverly Hills, CA: Sage.

Latour, B., & Woolgar, S. (1986). *Laboratory life: The social construction of scientific fact.* Princeton, NJ: Princeton University Press.

Levrini, O., Fantini, P., Tasquier, G., Pecori, B., & Levin, M. (2015). Defining and operationalizing appropriation for science learning. *Journal of the Learning Sciences, 24*(1), 93–136.

Lunsford, A. A., & Lunsford, K. J. (2008). "Mistakes are a fact of life": A national comparative study. *College Composition and Communication, 59*(4), 781–806.

Lunsford, E., Melear, C. T., Roth, W. M., Perkins, M., & Hickok, L. G. (2007). Proliferation of inscriptions and transformations among preservice science teachers engaged in authentic science. *Journal of Research in Science Teaching, 44*(4), 538–564.

Machina, H., & Wild, D. (2013). Electronic laboratory notebooks: Progress and challenges in implementation. *Journal of Laboratory Automation, 18*(4), 264–268.

McGonigal, J. (2011). *Reality is broken: Why games make us better and how they can change the world.* New York, NY: Penguin Press.

Morrison, J. (2008). Elementary preservice teachers' use of science notebooks. *Journal of Elementary Science Education, 20*(2), 13–21.

Muir, H. (2008). Science of Mentos–Diet Coke explosions explained. *New Scientist.* Retrieved from newscientist.com/article/dn14114-science-of-mentos-diet-coke-explosions-explained/

National Research Council. (2012). *A framework for K–12 science education: Practices, cross-cutting concepts and core ideas.* Washington, DC: National Academies Press.

Nave, R. (n.d.). Principal focal length. Retrieved from hyperphysics.phy-astr.gsu.edu/hbase/geoopt/foclen.html

Newton, I. (1671/1672). A letter of Mr. Isaac Newton, professor of the mathematicks in the University of Cambridge; containing his new theory about light and colors: Sent by the author to the publisher from Cambridge, Feb. 6. 1671/72; In order to be communicated to the R. Society. *Philosophical Transactions of the Royal Society, 6*, 3075–3087.

NGSS Lead States. (2013). *Next Generation Science Standards: For states, by states.* Washington, DC: National Academies Press.

Otero, V., Pollock, S., & Finkelstein, N. (2010). A physics department's role in preparing physics teachers: The Colorado learning assistant model. *American Journal of Physics, 78*(11), 1218.

Petraglia, J. (1995). Writing as an unnatural act. In J. Petraglia (Ed.), *Reconceiving writing, rethinking writing instruction* (pp. 79–100). Mahwah, NJ: Erlbaum.

Pozzer-Ardenghi, L., & Roth, W.-M. (2010). Toward a social practice perspective on the work of reading inscriptions in science texts. *Reading Psychology, 31*, 228–253.

Rathjen, D., & Doherty, P. (2002). *Square wheels and other easy-to-build hands-on science activities. An Exploratorium science snackbook.* San Francisco, CA: Exploratorium.

Roberson, C., & Lankford, D. (2010). Laboratory notebooks in the science classroom. *Science Teacher, 77*(1), 38–42.

Rooksby, J., & Ikeya, N. (2012). Collaboration in formative design: Working together at a whiteboard. *IEEE Software, 29*(1), 56–60. doi:10.1109/MS.2011.123

Roth, W.-M., & McGinn, M. K. (1998). Inscriptions: Toward a theory of representing as social practice. *Review of Educational Research, 68*, 35–59.

Russell, D. (1991). *Writing in the academic disciplines, 1870–1990: A curricular history.* Carbondale: Southern Illinois University Press.

Russell, D. (1995). Activity theory and its implications for writing instruction. In J. Petraglia (Ed.), *Reconceiving writing, rethinking writing instruction* (pp. 51–77). Mahwah, NJ: Erlbaum.

Scribner, S., & Cole, M. (1981). *The psychology of literacy.* Cambridge, MA: Harvard University Press.

Shankar, K. (2007). Order from chaos: The poetics and pragmatics of scientific recordkeeping. *Journal of the American Society for Information Science and Technology, 58*(10), 1457–1466.

Shankar, K. (2009). Ambiguity and legitimate peripheral participation in the creation of scientific documents. *Journal of Documentation, 65*(1), 151–165.

Sharma, A., & Anderson, C. W. (2009). Recontextualization of science from lab to school: Implications for science literacy. *Science & Education, 18*, 1253–1275.

Shaughnessy, M. P. (1979). *Errors and expectations: A guide for the teacher of basic writing*. New York, NY: Oxford University Press.

Shipka, J. (2011). *Toward a composition made whole*. Pittsburgh, PA: University of Pittsburgh Press.

Shipka, J. (2014). Beyond text and talk: A multimodal approach to first-year composition. In D. Teague & R. Lunsford (Eds.), *First-year composition: From theory to practice* (pp. 211–235). Anderson, SC: Parlor Press.

Soja, E. (1996). *Thirdspace: Journeys to Los Angeles and other real-and-imagined places*. Oxford, UK: Basil Blackwell.

Street, B. V. (1984). *Literacy in theory and practice* (Vol. 9). Cambridge, UK: Cambridge University Press.

Suchman, L. A. (1988). Representing practice in cognitive science. *Human Studies, 11*(2/3), 305–325. Retrieved from jstor.org/stable/20009029

Weiser, M. E., Fehler, B., & González, A. M. (Eds.). (2009). *Engaging audience: Writing in an age of new literacies*. Urbana, IL: National Council of Teachers of English.

Whitin, P., & Whitin, D. J. (1997). *Inquiry at the window: Pursuing the wonders of learners*. Portsmouth, NH: Heinemann.

Wiley, M. (2000). The popularity of formulaic writing (and why we need to resist). *The English Journal, 90*(1), 61–67.

Index

About the Authors

Leslie Atkins Elliott is an associate professor in Curriculum, Instruction, and Foundational Studies at Boise State University, with research and teaching interests in science education and teacher preparation. She has a PhD in physics from the University of Maryland, where she bridged education research and a physics degree as she investigated students' and scientists' uses of analogy in constructing scientific ideas. She has taught high school physics, chemistry, and undergraduate physics. Her most recent research projects involve understanding how ideas from science classrooms are noticed, valued, and used in everyday life, and understanding what characteristics of instruction help initiate and sustain productive disciplinary engagement in science.

Kim Jaxon is an associate professor of English (Composition & Literacy) at California State University, Chico. She received her PhD at UC, Berkeley in the Language & Literacy, Society & Culture program in the Graduate School of Education. Her research interests focus on theories of literacy, particularly digital literacies, participation, classroom design, game theories, and teacher education. In her research and teaching, she uses a variety of digital platforms and considers the affordances in terms of student learning and participation. She has published a variety of book chapters and articles focused on classroom design, mentoring, and the uses of digital tools. Kim recently was awarded the Teacher of Excellence-College Award by the California Association of Teachers of English.

Irene Salter received a PhD in neuroscience from the University of California, San Francisco. Over the course of the following decade, she taught middle school science and math, led professional development workshops at the Exploratorium Teacher's Institute, developed curriculum with the GEMS group at the Lawrence Hall of Science, and taught science to preservice teachers at California State University, Chico, where she served as the chair of the Department of Science Education. These experiences are now intertwined in her role as the principal of Chrysalis Charter School, a science- and nature-focused K–8 school near Redding, California.